# Modelling and simulation of inelastic phenomena in the material behaviour of steel during heat treatment processes

von Simone Bökenheide

Dissertation
zur Erlangung des Grades eines Doktors der Naturwissenschaften
– Dr. rer. nat. –

Vorgelegt im Fachbereich 3 (Mathematik & Informatik)
der Universität Bremen
im Juni 2015

Datum des Promotionskolloquiums: 10. Juli 2015

Gutachter: PD Dr. Michael Wolff (Universität Bremen) und
          Prof. Dr. Alfred Schmidt (Universität Bremen)

Bibliographic information published by the Deutsche Nationalbibliothek

The Deutsche Nationalbibliothek lists this publication in the Deutsche
Nationalbibliografie; detailed bibliographic data are available
on the Internet at http://dnb.d-nb.de .

ISBN 978-3-8325-4099-9

Logos Verlag Berlin GmbH
Comeniushof, Gubener Str. 47,
10243 Berlin
Tel.: +49 (0)30 42 85 10 90
Fax: +49 (0)30 42 85 10 92
INTERNET: http://www.logos-verlag.de

# Abstract

This work deals with the mathematical modelling of material behaviour of steel during heat treatment. Considering the material behaviour, we distinguish between elastic and inelastic phenomena. In the case of inelastic material behaviour, the underlying system of equations is non-linear and its equations are coupled. Therefore, the description and modelling of inelastic phenomena is especially challenging. During heating and austenitisation, workpieces are exposed to high temperatures during long time periods. This can - even under moderate stresses - lead to creep and thus to a distortion of the workpiece. Moreover, phase transformations in steel under non-zero deviatoric stresses cause transformation-induced plasticity (TRIP). The aim of this thesis is to describe, to model as well as to implement the material behaviour of steel during heating and austenitisation taking into account the above described phenomena.

We obtain a system of partial and ordinary differential equations for temperature, phase fractions and mechanical deformations. We deal with the solving of the system as well as with the implementation of the model equations. A numerical algorithm is developed in order to solve the coupled system of equations involving the inelastic quantities. As inelastic phenomena, we focus especially on creep and transformation-induced plasticity. In order to determine a specific material behaviour, the verification of concrete material laws is necessary. Furthermore, the knowledge of certain material parameters is required. We develop a procedure for the verification of material laws and for the identification of parameters. We use the obtained material parameters for simulations of three-dimensional material behaviour.

First, we consider the discretisation of the one-dimensional version of the model. We develop a procedure for the verification of material laws and for the identification of parameters. By the identified material parameters and the discretised model equations, we are able to implement the model equations and to perform simulations in a 1D as well as in a 3D setting. This enables us to conduct 3D simulations of the heat treatment of a workpiece under realistic conditions.

We study the material behaviour during different heat treatment scenarios. The simulations cover temperature, phase fractions and mechanical deformations including the inelastic effects creep and TRIP. The implementation of the model equations was carried out with the Finite Element Toolbox ALBERTA. We validate the 3D model by means of experimental data from workpiece experiments.

Another important part of this work covers the topic of multi-mechanism models. This work was developed as part of the research project BO114/4-1 'Multi-mechanism models: Theory and their application to some phenomena in material behaviour of steel' supported by the German Research Foundation (DFG). The idea of this approach is to decompose the inelastic part of the total strain into several parts, also referred to as mechanisms. Each of them can be studied separately with an own material law. In contrast to rheological models, it is possible for the mechanisms to interact with each other. We develop a two-mechanism model for creep and TRIP arising simultaneously and discuss thermodynamic consistency.

## Zusammenfassung

Diese Arbeit befasst sich mit der mathematischen Modellierung des Materialverhaltens von Stahl. Hierbei wird zwischen elastischem und inelastischem Verhalten unterschieden. Insbesondere die Beschreibung und Modellierung bestimmten inelastischen Verhaltens stellt eine Herausforderung dar, da zugrunde liegende Gleichungssysteme aus nichtlinearen, gekoppelten Gleichungen bestehen. Während der Erwärmung und Austenitisierung innerhalb der Wärmebehandlung werden Stahlbauteile über lange Zeiten hohen Temperaturen ausgesetzt. Dies kann bereits unter kleinen Spannungen zu inelastischer Verzerrung, genauer gesagt zu Kriechen führen. Das Ziel dieser Arbeit ist es, die Erwärmphase und die dort auftretenden Effekte im Material zu beschreiben, zu modellieren sowie zu implementieren.

In unserem Falle erhält man ein gekoppeltes nichtlineares Rand- Anfangswertproblem aus gewöhnlichen und partiellen Differentialgleichungen. Der Kern dieser Arbeit beschäftigt sich mit der Lösung des Systems sowie der Implementierung der Modellgleichungen. Hierbei wird ein numerischer Algorithmus entwickelt um das gekoppelte Gleichungssystem der inelastischen Größen zu lösen. Die betrachteten inelastischen Phänomene sind insbesondere Kriechen und Umwandlungsplastizität.

Um ein Materialverhalten zu bestimmen, müssen die zugrunde liegenden Materialgesetze verifiziert werden. Des Weiteren werden für die zugrunde liegenden Materialgestze bestimmte Materialparameter benötigt; diese werden a-priori mit Hilfe einer Parameteridentifikation bestimmt. Die so erhaltenen Parameter können in das 3D Modell eingesetzt werden. Hierdurch wird ermöglicht, dreidimensionale Simulationen durchzuführen.

Das dreidimensionale Problem wird zunächst auf den eindimensionalen Fall reduziert und in diskretisierter Form betrachtet. Mithilfe experimenteller Daten wird eine Parameteridentifikation durchgeführt und die simulierten mit den experimentellen Daten verifiziert. Die erhaltenen Parameter können nun in das implementierte 3D Simulationsmodell eingefügt werden. Dies ermöglicht uns dreidimensionale Simulationen durchzuführen und somit realitätsnahe Ergebnisse zu erhalten. Es werden unterschiedliche Szenarien bei der Wärmebehandlung eines Werkstücks betrachtet. Hierbei werden Temperatur, Phasenumwandlungen und Deformation des Werkstücks sowie inelastisches Materialverhalten durch Kriechen und Umwandlungsplastizität einbezogen. Die Implementierung erfolgt in dem Finite Elemente Tool ALBERTA. Das 3D Modell wird mit Hilfe von Experimentaldaten aus Bauteilversuchen validiert.

Ein weiterer Teil dieser Arbeit umfasst das Thema der Mehr-Mechanismen Modelle. Die Idee dieses Modellierungsansatzes ist es, den Anteil der inelastischen Verzerrung der Gesamtverzerrung in mehrere Komponenten zu zerlegen (sogenannte Mechanismen). Diese können jeweils einzeln mit eigenem Materialgesetz betrachtet werden. Hierbei besteht - im Gegensatz zu rheologischen Modellen - die Möglichkeit der Wechselwirkung der einzelnen Mechanismen. Insbesondere wurde das simultane Auftreten von Kriechen und Umwandlungsplastizität betrachtet, die Materialgesetze für ein Modell mit zwei Mechanismen aufgestellt sowie auf thermodynamische Konsistenz überprüft.

# Acknowledgements

First of all, my sincere gratitude goes to PD Dr. Michael Wolff for his inspirational supervision of this thesis, his courteous support and advice as well as many fruitful discussions. I also thank Prof. Dr. Michael Böhm for his guidance and advice. I would like to acknowledge Prof. Dr. Alfred Schmidt for his interest in my work and for his support in questions concerning numerical mathematics. Furthermore, I thank especially Dr. Jonathan Montalvo Urquizo and Dr. Bettina Suhr for a lot of helpful advices and many fruitful discussions both mathematically and personal.

I acknowledge the German Research Foundation (DFG) for the funding of my work via the research projects BO114/4-1 'Multi-mechanism models: Theory and their application to some phenomena in material behaviour of steel' as well as the Collaborative Research Centre SFB 570 'Distortion Engineering' at the University of Bremen and all colleagues in the projects.

I am grateful to the Stiftung Institut für Werkstofftechnik (IWT) in Bremen for providing material parameters and experimental data. Furthermore, I thank Dr. Münip Dalgic and Dr. Holger Surm from the Stiftung Institut für Werkstofftechnik (IWT) in Bremen for helpful discussions concerning experiments.

My special thanks goes to Dr. Sören Boettcher and Dr. Nils Hendrik Kröger for their constant support and for the careful reading of the manuscript providing many helpful hints and suggestions.

My deepest gratitude goes to the support of my family and all my colleagues and friends who encouraged me during writing this thesis, Dr. Iwona Piotrowska-Kurczewski, Dr. Dorota Kubalinska, Dr. Kamil Kazimierski, Dr. Christina Brandt, Eusevia Torrico Jiménez, Martina Unterländer and Mariana Altenburg Soppa for their constant belief and friendship and for reminding me to keep positive.

Simone Bökenheide

# Contents

# List of Figures

# List of Tables

# 1 Introduction

The main topic of this work is the mathematical modelling of material behaviour of steel during heat treatment. As an extensive used structural material worldwide, steel and steel components provide many applications which represent a very active field of research. Particularly, the prediction of distortion of a workpiece is of great interest.

Considering the material behaviour, we distinguish between elastic and inelastic phenomena. An elastic response of the material is reversible, whereas inelastic material behaviour leads to a permanent distortion. In order to be able to predict such distortional material behaviour, a detailed understanding of the arising phenomena is crucial.

In this work we deal with the investigation of phenomena arising during heat treatment of steel. Considering the heat treatment of steel, heating before hardening or case hardening plays an important role. During heating and austenitisation, workpieces are exposed to high temperatures over long time periods. This can - even under moderate stresses - lead to creep and thus to a distortion of the workpiece. Creep is a complex material behaviour occurring in mechanical structures under moderate stresses and relatively high temperatures. The analysis of creep as well as its simulation is of great importance in many industrial applications. Moreover, phase transformations in steel under non-zero deviatoric stresses yield a permanent volume-preserving deformation, even if the yield stress is not reached. This phenomenon is referred to as transformation-induced plasticity, abbreviated as TRIP. It is also possible that during heat treatment, creep and transformation-induced plasticity appear together.

This thesis was motivated by former investigations on processes involved in steel quenching (see Suhr (2010)). These investigations used the heated workpiece as initial situation for numerical simulations. Former effects arising during heating and austenitisation have not yet been taken into account. In particular, these are thermal effects, phase transformations as well as mechanical deformation including inelastic effects such as creep and transformation-induced plasticity. These phenomena can have an influence on the behaviour of the material afterwards. The aim of this thesis is bridging this gap by including the above described effects in the investigation.

Thereby, more realistic results considering the simulation of the material behaviour during the whole process, i.e. heating, austenitisation and quenching, can be expected.

From the mathematical point of view, we deal with a system of partial and ordinary differential equations for the temperature, phase fractions and mechanical deformations, see Chapter 2. In the case of inelastic material behaviour, the underlying system of equations is non-linear and its equations are coupled. Therefore, the description and modelling of inelastic phenomena is especially challenging. The presented models form the basis for the development of numerical algorithms for simulation and parameter identification. We will present the discretisation of the complete coupled model for temperature, phase fractions and mechanical deformation in Chapter 5. Here, a numerical algorithm is developed in order to solve the coupled system of equations involving the inelastic quantities, i.e. creep strain, back stress as well as stress, by an implicit solution scheme. This enables us to conduct 3D simulations of the heat treatment of a workpiece under realistic conditions.

In order to determine a specific material behaviour, the verification of concrete material laws is necessary. Furthermore, the knowledge of certain material parameters is required. In Chapter 4, we develop a procedure for the verification of possible laws for creep and TRIP and for the identification of parameters (cf. Wolff et al. (2012c)). The necessary data for the parameter identification are obtained by uniaxial experiments with special testing devices. We use the obtained material parameters for simulations of three-dimensional material behaviour.

Altogether, by means of the identified material parameters and the discretised model equations, we are able to implement the model equations and to perform simulations in a 1D as well as in a 3D setting.

In Chapter 6, we present results from one-dimensional simulations as well as from three-dimensional simulations of different workpieces.

In particular, we show simulations of uniaxial experiments for creep as well as for creep and TRIP during heating and austenitisation. Moreover, we present results of 3D simulations using the implemented developed numerical solution scheme. The simulations of material behaviour of different workpieces include thermal effects, mechanical deformation, creep as well as phase transformations. Finally, we validate the 3D model by means of experimental data from workpiece experiments.

We study the material behaviour during different heat treatment scenarios. The simulations cover temperature, phase fractions and mechanical deformations including the inelastic effects creep and TRIP. The implementation of the 3D model equations was carried out with the Finite Element Toolbox ALBERTA which was developed by Schmidt and Siebert (2005). As this software is open source, there are less restrictions than in commercial Finite Element tools and allows for full flexibility to implement our models with all their specifications considering equations, material parameters and numerical schemes for the spatial and temporal Finite Element discretisations.

Another important part of this work is the modelling by multi-mechanism models which we deal with in Chapter 3. These models represent a special approach for the modelling of inelastic material behaviour and have become an important tool for modelling complex material behaviour (see Saï (2011) for an overview). In the case of small deformations, the inelastic part of the total strain is decomposed into several parts (also referred to as mechanisms). We give a general introduction into the theory and the application of multi-mechanism models. Furthermore, we consider additional inelastic phenomena besides creep and TRIP: dislocation-based plasticity (to which we refer to as classical plasticity) as well as viscoplasticity. We distinguish these phenomena from creep.

We will focus especially on models with two mechanisms. In particular, we develop a two-mechanism model for creep as well as for creep and transformation-induced plasticity arising simultaneously. To our knowledge, the latter case has not yet been investigated. We provide both the corresponding material laws and discuss thermodynamic consistency of the models. This work was developed as part of the research project BO114/4-1 'Multi-mechanism models: Theory and their application to some phenomena in material behaviour of steel' supported by the German Research Foundation (DFG).

This work is organised as follows:

- Chapter 2 provides elements of continuum mechanics and thermodynamics. We present the required balance equations and material laws. We close the chapter with the presentation of the complete coupled model considering temperature, phase fractions and mechanical deformation.

- In Chapter 3, we focus on multi-mechanism models. We present the application of some two-mechanism models to different material behaviour and discuss thermodynamic consistency.

- Chapter 4 provides the 1D model equations. We describe the numerical algorithm to solve the problem and to verify certain material laws. Furthermore, we deal with parameter identification.

- Chapter 5 handles the discretisation of the three-dimensional model. We focus on weak formulation and on the application of the Finite Element method. We develop a solution scheme for the calculation of the inelastic quantities.

- In Chapter 6, we present both results from 1D simulations and from FEM simulations of some three-dimensional workpieces of steel. Furthermore, we show results from experimental data.

- Finally, we conclude this thesis with final remarks and future prospects in Chapter 7.

Further references to the different topics will be given in the specific chapters.

# 2 Continuum mechanics and theory of inelastic material behaviour

In this chapter, we will provide elements of continuum mechanics and thermodynamics. We will present the required balance equations and the conditions for a thermodynamic consistent model. First, we will provide balance equations and a special form of the heat conduction equation (see Section 2.1). In this work, we will especially focus on material behaviour of steel. Section 2.2 focuses on inelastic material behaviour. We will consider different kinds of inelastic material behaviour of steel and present the corresponding material laws.

Besides thermal effects described by the heat conduction equation, we also have to take phase transformations as well as mechanical behaviour into account.

Finally, Section 2.4 will present the complete coupled macroscopic model describing material behaviour of steel including specific inelastic phenomena and phase transformations.

A general overview of continuum mechanics and thermodynamics is given in Altenbach and Altenbach (1994), Altenbach (2012), Lemaitre and Chaboche (1990), Besson et al. (2001), Haupt (2002), Betten (2001) and Wolff et al. (2008). Details about elasticity and elastic materials are provided in Feynman et al. (1991). For more information about inelastic material behaviour, we refer to Simo and Hughes (1998), de Souza Neto et al. (2008), e.g.

For additional information about phase transformation we refer to Porter and Easterling (1992). A general introduction into materials sciences can be found in Berns and Theisen (2008), Ilschner and Singer (2005), Horstmann (1992) and Seidel (1999), e.g. Different approaches for the modelling of phase transformations in steel can be found in Wolff et al. (2007).

## 2.1 Continuum mechanical preparations

In the following, we consider the space-time domain $\Omega \times (0, T)$ where the presented relations have to be fulfilled. Here, $\Omega \subset \mathbb{R}^3$ denotes a Lipschitz-bounded domain and $T > 0$ stands for the process time. The quantities considered in the following, i.e. temperature, phase fractions, deformation (as well as material parameters) are depending on space and time $t$. We will suppress the dependence on $x, t$ for a better readability, for instance $\theta = \theta(x, t)$ denotes the temperature at the point $x \in \Omega$ at time $t$.

## 2.1.1 Balance equations

We restrict ourselves to small deformations. We set $\Omega$ to be the reference configuration and define the displacement vector $\boldsymbol{u} : \bar{\Omega} \times (0, T) \to \mathbb{R}^3$ as well as the strain tensor $\boldsymbol{\varepsilon} : \bar{\Omega} \times (0, T) \to \mathbb{R}^{3 \times 3}$. First, we provide the balance equations which have to be fulfilled in the space-time domain $\Omega \times (0, T)$: the equation of momentum, the energy equation and the Clausius-Duhem inequality:

$$\varrho_0 \, \ddot{\boldsymbol{u}} - \operatorname{div} \boldsymbol{\sigma} = \varrho_0 \boldsymbol{f} \qquad (2.1.1)$$

$$\varrho_0 \, \dot{e} + \operatorname{div} \boldsymbol{q} = \boldsymbol{\sigma} : \dot{\varepsilon} + r_\theta \qquad (2.1.2)$$

$$-\varrho_0 \, \dot{\psi} - \varrho_0 \, \eta \, \dot{\theta} + \boldsymbol{\sigma} : \dot{\varepsilon} - \frac{1}{\theta} \, \boldsymbol{q} \cdot \nabla \theta \; \geq \; 0 \; . \qquad (2.1.3)$$

Here, $\varrho_0$ represents the density in the reference configuration (i.e. at $t = 0$), $\boldsymbol{u}$ denotes the displacement vector, $\boldsymbol{\sigma} : \bar{\Omega} \times (0, T) \to \mathbb{R}^{3 \times 3}$ the (symmetric) Cauchy stress tensor, $\boldsymbol{f}$ represents the mass density of external forces.

In (2.1.2), $e$ stands for the mass density of the internal energy, $\boldsymbol{q}$ represents the heat-flux density vector, $\boldsymbol{\varepsilon}$ the linearised Green strain tensor and $r_\theta$ stands for the volume density of the heat supply. Here, $\boldsymbol{\alpha} : \boldsymbol{\beta}$ denotes the scalar product of two second-order tensors $\boldsymbol{\alpha}$ and $\boldsymbol{\beta}$, i.e. $\boldsymbol{\alpha} : \boldsymbol{\beta} := \sum_{i,j=1}^n \alpha_{ij} \beta_{ij} = \operatorname{tr}(\boldsymbol{\alpha} \boldsymbol{\beta}^T)$ .

The Clausius-Duhem inequality (2.1.3) ensures the thermodynamical consistency of the model. Here, $\psi$ is the mass density of free (or Helmholtz) energy, $\eta$ denotes the mass density of entropy and $\theta = \theta(x, t)$ the temperature.

We use the definition of the total strain tensor $\boldsymbol{\varepsilon}$:

$$\boldsymbol{\varepsilon}(\boldsymbol{u}) = \frac{1}{2} \left( \nabla \boldsymbol{u} + (\nabla \boldsymbol{u})^T \right) \; . \qquad (2.1.4)$$

The total strain tensor is split into a thermoelastic and an inelastic part, i.e.

$$\boldsymbol{\varepsilon} = \boldsymbol{\varepsilon}_{te} + \boldsymbol{\varepsilon}_{in} \; , \qquad (2.1.5)$$

where

$$\operatorname{tr}(\boldsymbol{\varepsilon}_{in}) = 0 \; . \qquad (2.1.6)$$

One can also suppose a further splitting of the inelastic strain tensor $\boldsymbol{\varepsilon}_{in}$ (cf. Section 3.1.1). Details will be given in Chapter 3.

## 2.1.2 Thermodynamics with internal variables

For the free energy $\psi$ in (2.1.3), we assume that $\psi$ is a function depending on temperature $\theta$, the thermoelastic strain part $\boldsymbol{\varepsilon}_{te}$ as well as on (tensorial or scalar) internal variables. The internal variables will be denoted by $\xi$. Altogether, we have $\psi = \psi(\boldsymbol{\varepsilon}_{te}, \xi, \theta)$.

Considering inelastic material behaviour, we propose a split of the free energy $\psi$ into a thermoelastic and an inelastic part. We decompose $\psi$ into

$$\psi = \psi(\boldsymbol{\varepsilon}_{te}, \xi, \theta) = \psi_{te}(\boldsymbol{\varepsilon}_{te}, \theta) + \psi_{in}(\xi, \theta) \; . \qquad (2.1.7)$$

In the case of isotropy, the thermoelastic part $\psi_{te} = \psi_{te}(\boldsymbol{\varepsilon}_{te}, \theta)$ of $\psi$ is usually given by

$$\psi_{te} := \frac{1}{\varrho_0}\left(\mu\,\boldsymbol{\varepsilon}_{te} : \boldsymbol{\varepsilon}_{te} + \frac{\lambda}{2}(\mathrm{tr}(\boldsymbol{\varepsilon}_{te}))^2 - 3K\alpha_\theta(\theta - \theta_0)\,\mathrm{tr}(\boldsymbol{\varepsilon}_{te}) + \frac{9}{2}K\alpha_\theta^2(\theta - \theta_0)^2\right) + C(\theta) \tag{2.1.8}$$

where $K$ stands for the compression modulus, $\alpha_\theta$ for the linear coefficient of thermal expansion, $C$ is the calorimetric function and $\lambda$, $\mu$ stand for the Lamé coefficients (see Wolff et al. (2008), Wolff et al. (2010) and Helm and Haupt (2003), e.g.).

The inelastic part $\psi_{in}$ of the free energy is assumed to be dependent on temperature and the internal variables $\xi$,

$$\psi_{in} = \psi_{in}(\xi, \theta) \quad , \quad \text{with} \quad \xi = (\xi_1, \ldots, \xi_{N_\xi}) \, . \tag{2.1.9}$$

$N_\xi$ denotes the number of internal variables and $\xi_i$ can stand for a tensorial or scalar variable ($i = 1, \ldots, N_\xi$). In the following, we will specify the internal variables and give a definition for $\psi_{in}$ according to the specific material behaviour, see Section 2.2.

Moreover, the internal variables have to fulfil evolution equations which are usually ordinary differential equations with respect to time (cf. Wolff et al. (2010), Wolff et al. (2011c)).

We use the following form of evolution equations:

$$\dot{\xi}_i = f_{\xi_i}(\boldsymbol{\sigma}, \xi, \dot{\xi}, \theta, \dot{\theta}, \dot{\boldsymbol{\varepsilon}}) \quad , \qquad \text{for} \quad i = 1, \ldots, N_\xi \, , \tag{2.1.10}$$

(see Wolff et al. (2015) for further explanations).

We set zero initial conditions:

$$\xi_i(0) = 0 \quad , \qquad \text{for} \quad i = 1, \ldots, N_\xi \, . \tag{2.1.11}$$

Using standard arguments of thermodynamics (cf. Lemaitre and Chaboche (1990), Maugin (1992), Wolff et al. (2010), Wolff et al. (2011c) and the references therein), the Clausius-Duhem inequality (2.1.3) can be simplified.

By (2.1.5) and (2.1.7), we can rewrite (2.1.3) as follows:

$$-\varrho_0\,\dot{\theta}\left(\frac{\partial\psi}{\partial\theta} + \eta\right) - \varrho_0\sum_{j=1}^{N_\xi}\frac{\partial\psi}{\partial\xi_j} : \dot{\xi}_j + \left(\boldsymbol{\sigma} - \varrho_0\frac{\partial\psi}{\partial\boldsymbol{\varepsilon}_{te}}\right) : \dot{\boldsymbol{\varepsilon}}_{te} + \boldsymbol{\sigma} : \dot{\boldsymbol{\varepsilon}}_{in} - \frac{1}{\theta}\,\boldsymbol{q}\cdot\nabla\theta \geq 0 \, . \tag{2.1.12}$$

Assuming the potential relations,

$$\boldsymbol{\sigma} = \varrho_0\frac{\partial\psi}{\partial\boldsymbol{\varepsilon}_{te}} \, , \tag{2.1.13}$$

and

$$\eta = -\frac{\partial\psi}{\partial\theta} \, , \tag{2.1.14}$$

the inequality (2.1.12) can be reduced to the *dissipation inequality*:

$$\boldsymbol{\sigma} : \dot{\boldsymbol{\varepsilon}}_{in} - \varrho_0 \sum_{j=1}^{N_\xi} \frac{\partial \psi}{\partial \xi_j} : \dot{\xi}_j - \frac{1}{\theta} \boldsymbol{q} \cdot \nabla \theta \geq 0 \,. \qquad (2.1.15)$$

A model is regarded as thermodynamically consistent, if the dissipation inequality (2.1.15) holds for all possible processes.

Assuming the Fourier law of heat conduction

$$\boldsymbol{q} = -k_\theta \, \nabla \theta \,, \qquad (2.1.16)$$

with the heat conductivity $k_\theta > 0$, the heat-conduction inequality

$$-\frac{1}{\theta} \boldsymbol{q} \cdot \nabla \theta \geq 0 \qquad (2.1.17)$$

is always fulfilled.

Furthermore, we define the thermodynamic forces:

$$\boldsymbol{\chi}_j = \varrho_0 \frac{\partial \psi}{\partial \xi_j} \,, \quad j = 1, \dots, N_\xi \,. \qquad (2.1.18)$$

By the preceding, it follows from (2.1.15) – (2.1.18) that our model is thermodynamically consistent if the *remaining inequality*

$$\boldsymbol{\sigma} : \dot{\boldsymbol{\varepsilon}}_{in} - \sum_{j=1}^{N_\xi} \boldsymbol{\chi}_j : \dot{\xi}_j \geq 0 \,, \qquad (2.1.19)$$

holds.

Taking the relation (2.1.13) for $\boldsymbol{\sigma}$, i.e.

$$\boldsymbol{\sigma} = \varrho_0 \frac{\partial \psi_{te}}{\partial \boldsymbol{\varepsilon}_{te}} \,, \qquad (2.1.20)$$

and (2.1.7), (2.1.8) into account, we obtain the (isotropic) Duhamel-Neumann (i.e. generalised Hooke) relation of linear thermoelasticity

$$\boldsymbol{\sigma} = \varrho_0 \frac{\partial \psi_{te}}{\partial \boldsymbol{\varepsilon}_{te}} = 2\mu \, \boldsymbol{\varepsilon}_{te} + \lambda \operatorname{tr}(\boldsymbol{\varepsilon}_{te}) \boldsymbol{I} - 3K \alpha_\theta \,(\theta - \theta_0) \boldsymbol{I} \,, \qquad (2.1.21)$$

(cf. Section 2.1.5).

## 2.1.3 Phase transformations

In this work, we focus on the material behaviour of steel. In macroscopic modelling, steel is regarded as a mixture of different phase fractions. Under certain conditions, one phase may transform into one another.

We set $M_p$ as the number of phases in the material where $M_p \geq 2$. Let

$$p_1, \ldots, p_{M_p} \tag{2.1.22}$$

denote the different phase fractions. The volume fraction of the $i$-th phase is represented by $p_i = p_i(x, t)$ for $i = 1, \ldots, M_p$ (cf. Remark 2.1.1).

Each of the phase fractions have to fulfil the following balance and non-negativity relations:

$$\sum_{i=1}^{M_p} p_i = 1 \qquad \text{and} \qquad p_i \geq 0, \qquad \text{for } i = 1, \ldots, M_p. \tag{2.1.23}$$

We set $p := (p_1, \ldots, p_{M_p})$. Generally, the evolution of the phase fractions is described by ordinary differential equations:

$$\dot{p}_i(x, t) = f_{p_i}(x, t, p, \theta, \ldots) \quad \text{for} \quad i = 1, \ldots, M_p. \tag{2.1.24}$$

In the case of several phases, the density $\varrho$ is a function of $\theta$ and $p$. It is obtained by means of the densities $\varrho_i$ of the single phase fractions. We use the mixture rule

$$\varrho(\theta, p) = \sum_{i=1}^{M_p} p_i \varrho_i(\theta), \tag{2.1.25}$$

for the total density of the phase mixture (cf. e.g. Mahnken et al. (2012)).

**Remark 2.1.1** (Volume vs. mass phase fractions). *Here, we consider volume phase fractions. These are mostly preferred in materials sciences (see Wolff et al. (2012c), Wolff et al. (2007)). Generally, in continuum-mechanical modelling mostly mass phase fractions are in use as they are independent of temperature and deformation. In this case, instead of (2.1.25) the total density can be expressed by*

$$\frac{1}{\varrho(\theta, p^M)} = \sum_{i=1}^{M_p} \frac{p_i^M}{\varrho_i(\theta)}. \tag{2.1.26}$$

*Further information can be found in Mahnken et al. (2012), Wolff et al. (2012c), Wolff et al. (2003) and Wolff et al. (2007).*

In Chapter 6, we will consider a specific example of a heat treatment experiment. In Section 2.3 we will focus on the modelling of the occurring phase transformations in this specific material in detail.

For additional information about phase transformation we refer to Porter and Easterling (1992). A general introduction into materials sciences can be found in Berns and Theisen (2008), Ilschner and Singer (2005), Horstmann (1992) and Seidel (1999), e.g. A general introduction as well as different approaches for the modelling of phase transformations in steel can be found in Wolff et al. (2007). For the modelling of phase transformations and TRIP, see Wolff and Böhm (2002b) and Wolff and Böhm (2002a).

## 2.1.4 Heat equation

For details about the derivation of the heat equation see Besson et al. (2001), Haupt (2002), Wolff et al. (2008) and Mahnken et al. (2012). In the following, we sketch the derivation of the heat equation from the energy balance (2.1.2) which is given by

$$\varrho_0 \, \dot{e} + \operatorname{div} \boldsymbol{q} = \boldsymbol{\sigma} : \dot{\boldsymbol{\varepsilon}} + r_\theta \ . \tag{2.1.27}$$

We use the Legendre transformation between the internal energy $e$ and the Helmholtz free energy $\psi$ which is given by

$$\psi = e - \theta\eta \ . \tag{2.1.28}$$

Assuming the entropy $\eta$ - and thus the internal energy $e$ - depend on the same variables as $\psi$ (see (2.1.7)), we have

$$\eta = \eta(\boldsymbol{\varepsilon}_{te}, \xi, \theta) \quad \text{and} \quad e = e(\boldsymbol{\varepsilon}_{te}, \xi, \theta) \ . \tag{2.1.29}$$

We consider the first part of (2.1.27). Using (2.1.28), this part can be rewritten as

$$\varrho_0 \, \dot{e} = \varrho_0 \, (\dot{\psi} + \dot{\theta}\eta + \theta\dot{\eta}) \ . \tag{2.1.30}$$

First, we express the first summand by the partial derivatives of $\psi$. Using the thermoelastic laws (2.1.13)–(2.1.14) and the definition of the thermodynamic forces (2.1.18), we obtain:

$$\begin{aligned}
\varrho_0 \, \dot{\psi} &= \varrho_0 \frac{\partial \psi}{\partial \boldsymbol{\varepsilon}_{te}} : \dot{\boldsymbol{\varepsilon}}_{te} + \varrho_0 \frac{\partial \psi}{\partial \xi} : \dot{\xi} + \varrho_0 \frac{\partial \psi}{\partial \theta} \dot{\theta} \\
&= \boldsymbol{\sigma} : \dot{\boldsymbol{\varepsilon}}_{te} + \sum_{j=1}^{N_\xi} \boldsymbol{\chi}_j : \dot{\xi}_j - \varrho_0 \eta \dot{\theta} \ .
\end{aligned} \tag{2.1.31}$$

In the same way, we handle the third part of the sum in (2.1.30). Taking the law (2.1.14) for $\eta$ into account, as well as (2.1.13)–(2.1.14), (2.1.18) and the definition of the specific heat

$$c_d := -\theta \frac{\partial^2 \psi}{\partial \theta^2} \ \left( = \frac{\partial e}{\partial \theta} \right) , \tag{2.1.32}$$

we obtain

$$\begin{aligned}
\varrho_0 \theta \dot{\eta} &= -\varrho_0 \theta \frac{\partial^2 \psi}{\partial \theta \partial t} \\
&= -\varrho_0 \theta \frac{\partial^2 \psi}{\partial \boldsymbol{\varepsilon}_{te} \partial \theta} : \dot{\boldsymbol{\varepsilon}}_{te} - \varrho_0 \theta \frac{\partial^2 \psi}{\partial \xi \partial \theta} : \dot{\xi} - \varrho_0 \theta \frac{\partial^2 \psi}{\partial \theta^2} \dot{\theta} \\
&= -\theta \frac{\partial \boldsymbol{\sigma}}{\partial \theta} : \dot{\boldsymbol{\varepsilon}}_{te} - \theta \sum_{j=1}^{N_\xi} \frac{\partial \boldsymbol{\chi}_j}{\partial \theta} : \dot{\xi}_j + \varrho_0 c_d \, \dot{\theta} \ .
\end{aligned} \tag{2.1.33}$$

Altogether, (2.1.27) with (2.1.30), (2.1.31) and (2.1.33) yield the following special form of the *heat conduction equation*:

$$\varrho_0 c_d \dot{\theta} - \mathrm{div}\,(k_\theta \nabla \theta) = \boldsymbol{\sigma} : \dot{\boldsymbol{\varepsilon}}_{in} - \sum_{j=1}^{N_\xi} \boldsymbol{\chi}_j : \dot{\xi}_j + \theta \frac{\partial \boldsymbol{\sigma}}{\partial \theta} : \dot{\boldsymbol{\varepsilon}}_{te} + \theta \sum_{j=1}^{N_\xi} \frac{\partial \boldsymbol{\chi}_j}{\partial \theta} : \dot{\xi}_j + r_\theta \,.$$
(2.1.34)

Here, we used (2.1.5) and the definition of the heat-flux vector $\boldsymbol{q}$ according to Fourrier's law (2.1.16).

The right-hand side of (2.1.34) represents the heat sources. The single parts refer to the mechanical dissipation, the thermoelastic dissipation as well as the dissipation due to the dependence of the thermomechanical forces on temperature. The term $r_\theta$ covers further contributions, i.e. thermal energy released or consumed for instance due to phase transformations (see following Subsection).

### Special form of heat conduction equation

In the case of phase transformations, there arise an additional inner heat source due to the so-called *latent heat*. In order to take phase transformations into account that arise during heating, we extend (2.1.34). The heat conduction equation is derived from the energy balance (2.1.27) where we use again (2.1.28).

Now, the free energy $\psi$ and the entropy $\eta$ additionally depend on the phase fractions $p = (p_1, \ldots, p_{M_p})$. Thus, for the internal energy $e$ we have

$$e = e(\boldsymbol{\varepsilon}_{te}, \xi, p, \theta) = \psi(\boldsymbol{\varepsilon}_{te}, \xi, p, \theta) + \theta \eta(\boldsymbol{\varepsilon}_{te}, \xi, p, \theta) \,,$$
(2.1.35)

(cf. (2.1.28)). We introduce the thermodynamic forces

$$Z_i := \varrho_0 \frac{\partial \psi}{\partial p_i}, \quad \text{for} \quad i = 1, \ldots, M_p \,,$$
(2.1.36)

and $Z = (Z_1, \ldots, Z_{M_p})$. Then, we consider the further partial derivatives with respect to $p$ in (2.1.27) using (2.1.30) and (2.1.35):

$$\varrho_0 \frac{\partial e}{\partial p}\, \dot{p} = \varrho_0 \frac{\partial \psi}{\partial p}\, \dot{p} + \varrho_0 \theta \frac{\partial \eta}{\partial p}\, \dot{p}$$
$$= Z\,\dot{p} - \varrho_0 \theta \frac{\partial^2 \psi}{\partial \theta\, \partial p}\, \dot{p} \,.$$
(2.1.37)

In the last step, we used (2.1.36) and (2.1.14). We make use of (2.1.36) again in the second part on the right-hand side of (2.1.37). Altogether, we have

$$\varrho_0 \frac{\partial e}{\partial p}\, \dot{p} = \left( Z - \theta \frac{\partial Z}{\partial \theta} \right) \dot{p} \,.$$
(2.1.38)

Due to the balance relation (2.1.23), it holds that

$$\sum_{i=2}^{M_p} \dot{p}_l = \dot{p}_1 \,,$$
(2.1.39)

and thus

$$Z\dot{p} = \sum_{i=1}^{M_p} Z_i \dot{p}_i = \sum_{i=2}^{M_p} Z_i \dot{p}_i + Z_1 \dot{p}_1$$

$$= \sum_{i=1}^{M_p} (Z_i - Z_1)\dot{p}_i \,, \tag{2.1.40}$$

as well as

$$\frac{\partial Z}{\partial \theta}\dot{p} = \sum_{i=1}^{M_p} \frac{\partial Z_i}{\partial \theta} \dot{p}_i = \sum_{i=1}^{M_p} \left( \frac{\partial Z_i}{\partial \theta} - \frac{\partial Z_1}{\partial \theta} \right) \dot{p}_i \,. \tag{2.1.41}$$

We insert (2.1.40) and (2.1.41) in (2.1.38) and obtain

$$\varrho_0 \frac{\partial e}{\partial p}\dot{p} = \left( Z - \theta \frac{\partial Z}{\partial \theta} \right) \dot{p} = \sum_{i=1}^{M_p} \left( Z_i - \theta \frac{\partial Z_i}{\partial \theta} \right) \dot{p}_i = \sum_{i=1}^{M_p} \left( Z_i - Z_1 - \frac{\partial Z_i}{\partial \theta} + \frac{\partial Z_1}{\partial \theta} \right) \dot{p}_i \,. \tag{2.1.42}$$

By introducing the latent heat

$$L_{i1} := - \left( Z_i - Z_1 - \frac{\partial Z_i}{\partial \theta} + \frac{\partial Z_1}{\partial \theta} \right) \,, \tag{2.1.43}$$

for the transformation $1 \to i$ or $i \to 1$, we can rewrite the heat conduction equation (2.1.34). We finally obtain the following form of the heat conduction equation:

$$\varrho_0 c_d \dot{\theta} - \operatorname{div}(k_\theta \nabla \theta) =$$

$$\boldsymbol{\sigma} : \dot{\boldsymbol{\varepsilon}}_{in} - \sum_{j=1}^{N_\xi} \boldsymbol{\chi}_j : \dot{\xi}_j + \theta \frac{\partial \boldsymbol{\sigma}}{\partial \theta} : \dot{\boldsymbol{\varepsilon}}_{te} + \theta \sum_{j=1}^{N_\xi} \frac{\partial \boldsymbol{\chi}_j}{\partial \theta} : \dot{\xi}_j + \sum_{i=1}^{M_p} L_{i1} \dot{p}_j + \hat{r}_\theta \,. \tag{2.1.44}$$

For further details, we refer to Besson et al. (2001), Mahnken et al. (2012) and Mahnken et al. (2015).

**Remark 2.1.2.** *An approach for the latent heat via enthalpy is presented in* Mahnken et al. (2015).

### Initial boundary value problem

In the following, we do not consider dissipation. Therefore, some terms in (2.1.44) can be neglected. We obtain the following special form of the heat-conduction equation together with initial and (mixed) boundary conditions:

$$\varrho_0 c_d \dot{\theta} - \operatorname{div}(k_\theta \nabla \theta) = \varrho_0 \sum_{i=2}^{M_p} L_i \dot{p}_i \qquad \text{in} \quad \Omega \times (0,T) \tag{2.1.45a}$$

$$\theta(x,0) = \theta_0 \qquad \text{in} \quad \Omega \tag{2.1.45b}$$

$$\theta(x,t) = \theta_{ext_D}(t) \qquad \text{on} \quad \partial\Omega \times (0,T_1) \tag{2.1.45c}$$

$$-k_\theta(\theta,p)\nabla\theta \cdot n = \delta(\theta(x,t) - \theta_{ext_R}(t)) \qquad \text{on} \quad \partial\Omega \times (T_1,T) \tag{2.1.45d}$$

where $T_1 \in [0, T]$, $k_\theta = k_\theta(\theta, p)$ stands for the heat conductivity and $L_i$ denotes the latent heats for the transformation into the forming phase, cf. (2.1.43), (2.1.44) (for details see Wolff et al. (2008) and Mahnken et al. (2012)). In (2.1.45d), $\delta$ represents the heat transfer coefficient. The initial temperature is $\theta_0$. The heat transfer at the boundary of the considered domain can be modelled by means of Dirichlet or Robin boundary conditions, see (2.1.45c) and (2.1.45d), respectively. Here, also the cases $T_1 = 0$ or $T_1 = T$ are possible. In (2.1.45c), (2.1.45d), $\theta_{ext}$ stands for the external temperature which can e.g. be taken from experimental data.

**Remark 2.1.3** (Boundary conditions).
*In the simulation results which will be presented in Chapter 6, the model was implemented in order to perform a 3D simulation of a workpiece. We used Dirichlet boundary conditions where the external temperature corresponds to the surface temperature of the workpiece. Considering quenching, the temperature of the cooling fluid can be used as $\theta_{ext}$ (see Suhr (2010)).*

## 2.1.5 Mechanics

We formulate the equation of linear momentum (see (2.1.1)) as initial boundary value problem:

$$\varrho_0\,\ddot{\boldsymbol{u}} - \operatorname{div}\boldsymbol{\sigma} = \varrho_0\boldsymbol{f} \qquad\qquad \text{in}\quad \Omega \times (0, T) \qquad (2.1.46\text{a})$$
$$\boldsymbol{u}(x, 0) = 0 \quad , \quad \dot{\boldsymbol{u}}(x, 0) = 0, \qquad\qquad \text{in}\quad \Omega \qquad (2.1.46\text{b})$$
$$\boldsymbol{u}(x, t) = g_D \qquad\qquad \text{on}\quad \Gamma_D \times (0, T) \qquad (2.1.46\text{c})$$
$$\boldsymbol{\sigma} \cdot n = g_N \qquad\qquad \text{on}\quad \Gamma_N \times (0, T) \qquad (2.1.46\text{d})$$

where $\partial\Omega = \Gamma_D \cup \Gamma_N$. The Neumann boundary condition in (2.1.46d) can be used to apply an external load $g_N$. The inertia term, i.e. the first term in (2.1.46), is often neglected.

The presented partial differential equation (2.1.46) is of hyperbolic type. If the inertia term is neglected, the deformation equation is a partial differential equation of elliptic type.

**Elasticity**

A general introduction to elastic material behaviour can be found e.g. in Bertram and Glüge (2013), Haupt (2002).

in the case of linear elasticity, Hooke's law states that

$$\boldsymbol{\sigma} = \boldsymbol{C}\boldsymbol{\varepsilon}, \qquad (2.1.47)$$

where $\boldsymbol{C}$ stands for the elasticity tensor. For isotropic materials, the elasticity tensor can be expressed by the Lamé coefficients $\lambda$ and $\mu$ which gives us

$$\boldsymbol{\sigma} = 2\mu\boldsymbol{\varepsilon} + \lambda(\operatorname{tr}(\boldsymbol{\varepsilon}))\boldsymbol{I}. \qquad (2.1.48)$$

For the non-isothermal case one obtains

$$\boldsymbol{\sigma} = 2\mu\boldsymbol{\varepsilon} + \lambda(\text{tr}(\boldsymbol{\varepsilon}))\boldsymbol{I} - 3K\alpha_\theta(\theta - \theta_0)\boldsymbol{I} \ , \tag{2.1.49}$$

cf. result in (2.1.21).

In the case of phase and temperature changes, the stress tensor is given by

$$\boldsymbol{\sigma} = 2\mu\boldsymbol{\varepsilon} + \lambda(\text{tr}(\boldsymbol{\varepsilon}))\boldsymbol{I} - 3K \left( \sqrt[3]{\frac{\varrho_0}{\varrho(\theta,p)}} - 1 \right) \boldsymbol{I} \ . \tag{2.1.50}$$

Here, $\lambda$ and $\mu$ denote the Lamé coefficients and $K$ stands for the compression modulus. Each depend on temperature and phase fractions. The last term in (2.1.50) covers density changes caused by changes of temperature as well as of phase fractions.

**Remark 2.1.4.** *The quantities $\lambda$, $\mu$ and $K$ can also be expressed in the following way by means of Young's modulus $E$ and Poisson's ratio $\nu$:*

$$\mu = \frac{E}{2(1+\nu)} \quad , \quad \lambda = \frac{\nu E}{(1+\nu)(1-2\nu)} \quad , \quad K = \lambda + \frac{2}{3}\mu = \frac{E}{3(1-2\nu)} \quad . \tag{2.1.51}$$

**Modelling of inelastic material behaviour**

In the general case of inelastic material behaviour (considering small deformations), the total strain $\boldsymbol{\varepsilon}$ is split into the thermoelastic strain $\boldsymbol{\varepsilon}_{te}$ and the inelastic strain $\boldsymbol{\varepsilon}_{in}$,

$$\boldsymbol{\varepsilon} = \boldsymbol{\varepsilon}_{te} + \boldsymbol{\varepsilon}_{in} \ . \tag{2.1.52}$$

where the inelastic strain is assumed to be traceless, i.e.

$$\text{tr}(\boldsymbol{\varepsilon}_{in}) = 0 \ . \tag{2.1.53}$$

The accumulated inelastic strain is defined as

$$s_{in}(t) := \int_0^t \sqrt{\frac{2}{3}\dot{\boldsymbol{\varepsilon}}_{in}(\tau) : \dot{\boldsymbol{\varepsilon}}_{in}(\tau)} \ \mathrm{d}\tau \ . \tag{2.1.54}$$

Generally, the inelastic strain $\boldsymbol{\varepsilon}_{in}$ can stand for different inelastic phenomena such as dislocation-based plasticity (to which we refer to as classical plasticity in the following), transformation induced plasticity (TRIP), viscoplasticity etc. In this work, we will focus especially on creep and TRIP. Besides this, also classical plasticity will be considered as the material laws will be required in the following Chapter 3. The material laws for the description of the corresponding inelastic strain will be handled in detail in Section 2.2.

**Remark 2.1.5.** *Additionally to the split in (2.1.52), the inelastic strain tensor $\boldsymbol{\varepsilon}_{in}$ can further be decomposed into several components, so-called 'mechanisms'. This yields more possibilities regarding the modelling of material behaviour. An introduction into the theory of multi-mechanism models together with the application to our problem will be presented in Chapter 3.*

In order to describe the stress tensor, we use the general assumption which states that the stress is determined by the elastic part of the strain (cf. (2.1.50), where $\varepsilon = \varepsilon_{el}$). Thus, we can describe the stress tensor by means of the elastic part of the total strain tensor in (2.1.52).

Using (2.1.50) together with (2.1.52) and (2.1.53), we obtain the following equation for the stress tensor:

$$\boldsymbol{\sigma} = 2\mu(\varepsilon - \varepsilon_{in}) + \lambda(\mathrm{tr}(\varepsilon))\boldsymbol{I} - 3K \left( \sqrt[3]{\frac{\varrho_0}{\varrho(\theta, p)}} - 1 \right) \boldsymbol{I} \ . \tag{2.1.55}$$

The following Chapter presents some examples of inelastic material behaviour together with the corresponding material laws.

## 2.2 Specific cases of inelastic material behaviour

In the following, we will present different examples of inelastic material behaviour which are of main interest in this work. We will present the material laws for the corresponding inelastic strains.

The first subsection 2.2.1 will present the specific material laws in the case of classical plasticity. After that, we will focus on creep behaviour, see Section 2.2.2. In Section 2.2.3, we will present the material laws for transformation induced plasticity.

Furthermore, in Subsection 2.2.4, we will briefly introduce the topic of multi-mechanism models.

An introduction and a description of different inelastic phenomena can be found in Haupt (2002), Simo and Hughes (1998), Betten (2001), Lemaitre and Chaboche (1990) e.g.

### 2.2.1 Classical plasticity

A general introduction into plastic material behaviour can be found in Han and Reddy (1999), Simo and Hughes (1998), de Souza Neto et al. (2008), e.g. Furthermore, we refer to Wolff et al. (2011b), Wolff et al. (2008) and Chaboche (2008).

In this section, we will specify the inelastic part of the free energy (2.1.7). We will provide the internal variables (cf. (2.1.9)) together with their evolution equations. We will obtain general approaches for *non-linear kinematic and isotropic hardening* in metal plasticity.

For the inelastic part of the free energy $\psi$ we set

$$\psi_{in} = \psi_{kin} + \psi_{iso} \ . \tag{2.2.1}$$

For the internal variables we set $\xi = (\boldsymbol{\alpha}, r)$, thus

$$\psi_{in} = \psi_{in}(\boldsymbol{\alpha}, r, \theta) = \psi_{kin}(\boldsymbol{\alpha}, \theta) + \psi_{iso}(r, \theta) \ . \tag{2.2.2}$$

The internal variables $\xi$ consist of a symmetric tensorial variable of strain type $\alpha$ related to kinematic hardening and one scalar internal variable $r$ related to isotropic hardening. For the single parts of $\psi_{in}$ we set

$$\psi_{kin}(\boldsymbol{\alpha}, \theta) := \frac{1}{3\varrho_0} c_{cp}(\theta) \boldsymbol{\alpha} : \boldsymbol{\alpha} \, , \qquad (2.2.3)$$

and

$$\psi_{iso}(r, \theta) := \frac{1}{2\varrho_0} d_{cp}(\theta) r^2 \, . \qquad (2.2.4)$$

Assuming

$$c_{cp} \geq 0 \quad \text{and} \quad d_{cp} \geq 0 \, , \qquad (2.2.5)$$

for all admissible temperatures $\theta$, the inelastic free energy $\psi_{in}$ is a convex function.

Next, we define the back stress $\boldsymbol{X}$ and the isotropic hardening $R$ as thermodynamic forces (see (2.1.18)) via the partial derivatives of the free energy with respect to the corresponding internal variables related to kinematic and isotropic hardening, respectively:

$$\boldsymbol{X} = \varrho_0 \frac{\partial \psi_{kin}}{\partial \boldsymbol{\alpha}} \quad \text{and} \quad R = \varrho_0 \frac{\partial \psi_{iso}}{\partial a} \, . \qquad (2.2.6)$$

Taking the ansatz (2.2.3) and (2.2.4) for the kinematic and isotropic parts of $\psi_{in}$ into account, we obtain:

$$\boldsymbol{X} = \frac{2}{3} c_{cp}(\theta) \boldsymbol{\alpha} \qquad (2.2.7)$$

and

$$R = d_{cp}(\theta) r \, . \qquad (2.2.8)$$

Furthermore, we assume the following general evolution law for the plastic strain $\boldsymbol{\varepsilon}_{cp}$:

$$\dot{\boldsymbol{\varepsilon}}_{cp} = \gamma \frac{\boldsymbol{\sigma}^* - \boldsymbol{X}^*}{\|\boldsymbol{\sigma}^* - \boldsymbol{X}^*\|} \, , \qquad (2.2.9)$$

where $\gamma \geq 0$ denotes the plastic multiplier. For a tensor $\boldsymbol{\tau}$ its deviator $\boldsymbol{\tau}^*$ is defined as

$$\boldsymbol{\tau}^* := \boldsymbol{\tau} - \frac{1}{3} \text{tr}(\boldsymbol{\tau}) \boldsymbol{I} \, . \qquad (2.2.10)$$

The norm $\|.\|$ is given by

$$\|\boldsymbol{\sigma}^*\| := \sqrt{\boldsymbol{\sigma}^* : \boldsymbol{\sigma}^*} \, . \qquad (2.2.11)$$

The accumulated plastic strain $s_{cp}$ is defined as

$$s_{cp}(t) := \int_0^t \sqrt{\frac{2}{3} \dot{\boldsymbol{\varepsilon}}_{cp}(\tau) : \dot{\boldsymbol{\varepsilon}}_{cp}(\tau)} \, \mathrm{d}\tau = \int_0^t \sqrt{\frac{2}{3}} \|\dot{\boldsymbol{\varepsilon}}_{cp}(\tau)\| \, \mathrm{d}\tau \, . \qquad (2.2.12)$$

By means of (2.2.9), we have:

$$\dot{s}_{cp} = \sqrt{\frac{2}{3}} \gamma \, . \qquad (2.2.13)$$

We define the yield function $f$ by

$$f(\boldsymbol{\sigma}, \boldsymbol{X}, R, R_0) := \sqrt{\frac{3}{2}} \|\boldsymbol{\sigma}^* - \boldsymbol{X}^*\| - (R_0 + R) , \qquad (2.2.14)$$

where $R_0$ denotes the initial yield stress. This function distinguishes the elastic from the plastic domain. We suppose the constraint

$$f(\boldsymbol{\sigma}, \boldsymbol{X}, R, R_0) \leq 0 . \qquad (2.2.15)$$

The plastic multiplier $\gamma$ has to fulfil

$$\gamma = 0 \quad \text{if} \quad f(\boldsymbol{\sigma}, \boldsymbol{X}, R, R_0) < 0 \qquad (2.2.16\text{a})$$
$$\gamma \geq 0 \quad \text{if} \quad f(\boldsymbol{\sigma}, \boldsymbol{X}, R, R_0) = 0 . \qquad (2.2.16\text{b})$$

In the next step, we assume an evolution equation for the internal variable $\boldsymbol{\alpha}$

$$\dot{\boldsymbol{\alpha}} = \dot{\boldsymbol{\varepsilon}}_{cp} - \frac{3}{2} \frac{a_{cp}}{c_{cp}} \boldsymbol{X} \dot{s}_{cp} , \qquad (2.2.17)$$

as well as for $r$ :

$$\dot{r} = \dot{s}_{cp} - \frac{b_{cp}}{d_{cp}} R \dot{s}_{cp} , \qquad (2.2.18)$$

where we require that

$$a_{cp}, b_{cp} \geq 0 \quad \text{and} \quad c_{cp}, d_{cp} > 0 . \qquad (2.2.19)$$

This leads to the integral equations

$$\boldsymbol{X}(t) = \frac{2}{3} c_{cp}(\theta) \left[ \boldsymbol{\varepsilon}_{cp} - \frac{3}{2} \int_0^t \frac{a_{cp}(\theta)}{c_{cp}(\theta)} \boldsymbol{X}(\tau) \dot{s}_{cp} \, \mathrm{d}\tau \right] \qquad (2.2.20)$$

$$R(t) = d_{cp}(\theta) \left[ s_{cp} - \int_0^t \frac{b_{cp}(\theta)}{d_{cp}(\theta)} R(s) \dot{s}_{cp} \, \mathrm{d}s \right] . \qquad (2.2.21)$$

Assuming *constant* $a_{cp}, b_{cp}, c_{cp}, d_{cp}$, we obtain the ordinary differential equations for the evolution of the back stress $\boldsymbol{X}$ and the isotropic hardening $R$

$$\dot{\boldsymbol{X}} = \frac{2}{3} c_{cp} \dot{\boldsymbol{\varepsilon}}_{cp} - a_{cp} \boldsymbol{X} \dot{s}_{cp} \qquad (2.2.22)$$

$$\dot{R} = d_{cp} \dot{s}_{cp} - b_{cp} R \dot{s}_{cp} , \qquad (2.2.23)$$

which represent the *generalised material law by Armstrong-Frederick* for non-linear hardening.

**Remark 2.2.1.** *In the case of temperature-depending coefficients, one obtains*

$$\dot{\boldsymbol{X}} = \frac{2}{3} c_{cp}(\theta) \dot{\boldsymbol{\varepsilon}}_{cp} - a_{cp}(\theta) \boldsymbol{X} \dot{s}_{cp} + \frac{2}{3} \frac{\dot{c}_{cp}(\theta)}{c_{cp}(\theta)} \boldsymbol{X} \qquad (2.2.24)$$

$$\dot{R} = d_{cp}(\theta) \dot{s}_{cp} - b_{cp}(\theta) R \dot{s}_{cp} + \frac{\dot{d}_{cp}(\theta)}{d_{cp}(\theta)} R , \qquad (2.2.25)$$

*using* (2.2.21), (2.2.7), (2.2.8), (2.2.17) *and* (2.2.18). *(Here, $\dot{c}_{cp}$ corresponds to the total time derivative of the function $c_{cp}(\theta(t))$.)*

## 2.2.2 Modelling of creep

Concerning general mechanisms and modelling of creep, we refer to Naumenko and Altenbach (2007),and Bökenheide and Wolff (2012), e.g. Betten (2002)

In the case of creep, we only consider kinematic hardening. We set $\xi = \alpha$ for the internal variables of the free energy $\psi$, thus

$$\psi_{in} = \psi_{in}(\alpha, \theta)\,, \tag{2.2.26}$$

where $\alpha$ denotes a symmetric tensorial variable of strain type related to kinematic hardening. The inelastic part of the free energy $\psi_{in}$ is defined as

$$\psi_{in}(\alpha, \theta) := \frac{1}{3\varrho_0} c_c(\theta) \alpha : \alpha\,, \tag{2.2.27}$$

Assuming

$$c_c \geq 0\,, \tag{2.2.28}$$

for all admissible temperatures $\theta$, the inelastic free energy $\psi_{in}$ is a convex function.

Next, we define the back stress $\boldsymbol{X}_c$ via the partial derivatives of the free energy with respect to $\alpha$, cf (2.1.18). This yields

$$\boldsymbol{X}_c = \varrho_0 \frac{\partial \psi_{in}}{\partial \alpha} = \frac{2}{3} c_c(\theta) \alpha \tag{2.2.29}$$

taking (2.2.27) into account.

Now, we specify the material law for the inelastic strain, that is the creep strain $\varepsilon_c$. We consider a commonly used model for the evolution of the creep strain $\varepsilon_c$,

$$\dot{\varepsilon}_c = \frac{3}{2} A \left( \sqrt{\frac{3}{2}} \frac{\|\sigma^* - \boldsymbol{X}_c^*\|}{D_c} \right)^{m-1} \frac{\sigma^* - \boldsymbol{X}_c^*}{D_c} s_c^k\,, \tag{2.2.30}$$

where $\sigma^*$ stands for the deviator of $\sigma$ and the norm is defined as in (2.2.11). The accumulated creep strain $s_c$ is defined as

$$s_c(t) := \int_0^t \sqrt{\frac{2}{3} \dot{\varepsilon}_c(\tau) : \dot{\varepsilon}_c(\tau)} \, \mathrm{d}\tau = \int_0^t \sqrt{\frac{2}{3}} \|\dot{\varepsilon}_c(\tau)\| \, \mathrm{d}\tau\,. \tag{2.2.31}$$

The material parameters $A > 0$, $m > 0$ and $k$ generally depend on temperature $\theta$ and possibly on further quantities like creep accumulation $s_c$. in the case of $\boldsymbol{X}_c \equiv 0$, the material function $k$ controls the stages of creep:

$$\begin{array}{ll} k < 0 & \text{primary creep,} \\ k = 0 & \text{secondary creep,} \\ k > 0 & \text{tertiary creep.} \end{array} \tag{2.2.32}$$

Thus, the approach in (2.2.30) covers all three creep stages. Often, in the case of tertiary creep, an evolution equation of a damage variable is added.

Next, we set an evolution equation for the internal variable:

$$\dot{\boldsymbol{\alpha}} = \dot{\boldsymbol{\varepsilon}}_c - \frac{3}{2}\frac{a_c}{c_c}\boldsymbol{X}_c(\dot{s}_c)^l, \tag{2.2.33}$$

Using (2.2.29) and (2.2.33), the evolution of $\boldsymbol{X}_c$ can be described by

$$\dot{\boldsymbol{X}}_c = \frac{2}{3}c_c(\theta)\,\dot{\boldsymbol{\varepsilon}}_c - a_c(\theta,\boldsymbol{\sigma})\,\boldsymbol{X}_c(\dot{s}_c)^l + \frac{\dot{c}_c(\theta)}{c_c(\theta)}\,\boldsymbol{X}_c, \tag{2.2.34}$$

with parameters $c_c > 0, a_c \geq 0$. The parameter $l \in \{0,1\}$ in (2.2.34) works as a switch: If $l = 1$, the rate law corresponds to the Armstrong-Frederick approach in plasticity (see model approach in Section 2.2.1). The case $l = 0$ yields an approach suggested by Robinson (cf. Arya and Kaufman (1989)). In (2.2.30), $D_c$ denotes the drag stress. In the simple case, it is supposed to be constant. Generally, one may assume an evolution equation in the form

$$\dot{D}_c = \gamma_c(\theta)\dot{s}_c - \beta_c(\theta,\boldsymbol{\sigma})\,D_c\,\dot{s}_c, \tag{2.2.35}$$

where $\gamma_c$ and $\beta_c$ represent material functions. Usually, one sets initial conditions as

$$\boldsymbol{X}_c(0) = 0, \qquad D_c(0) = 1. \tag{2.2.36}$$

The required material parameters in equations (2.2.30) and (2.2.34) have to be determined appropriately by means of experimental data. This parameter identification will be presented in detail in Section 4.6. The material parameters obtained by means of one-dimensional simulations will be used afterwards for the three-dimensional setting. This can be done due to the assumption of the equivalence of the uniaxial and multiaxial stress states, see Altenbach (2012).

More details about the algorithm to verify creep and TRIP behaviour using uniaxial experiments can be found in Bökenheide et al. (2012b) and Wolff et al. (2012c).

**Remark 2.2.2.** *In the case of uni-axial experiments with an applied stress $S$, the value of the norm is given by $\sqrt{\boldsymbol{\sigma}^* : \boldsymbol{\sigma}^*} = \sqrt{\frac{2}{3}}S$. The connection between the three-dimensional and the one-dimensional version of the model equation is the reason for the factor $\sqrt{\frac{3}{2}}$ in (2.2.30). For details about the equivalence of uniaxial stress states, see Altenbach (2012).*

## 2.2.3 Material laws for TRIP

For some general model approaches for transformation induced plasticity see for example Fischer et al. (1996), Leblond (1989), Wolff et al. (2009).

In the case of TRIP, we have one internal variable $\boldsymbol{\alpha}$ corresponding to kinematic hardening (cf. (2.1.9)). We define the inelastic part of the free energy $\psi_{in}$ in (2.1.7) as

$$\psi_{in}(\boldsymbol{\alpha},\theta) := \frac{1}{3\varrho_0}c_{trip}(\theta)\boldsymbol{\alpha} : \boldsymbol{\alpha}. \tag{2.2.37}$$

We assume that

$$c_{trip} \geq 0 , \tag{2.2.38}$$

for all admissible temperatures $\theta$, such that the inelastic free energy $\psi_{in}$ is a convex function.

Considering the TRIP strain $\varepsilon_{trip}$, we use the following approach for TRIP in the multi-phase case:

$$\dot{\varepsilon}_{trip} = \frac{3}{2} \sigma^* \sum_{i=1}^{M_p} \kappa_i \frac{d\phi_i}{dp_i} \max\{\dot{p}_i, 0\} . \tag{2.2.39}$$

Here, $\kappa_i > 0$ is the Greenwood-Johnson parameter (which may be temperature and stress-dependent) and $\phi_i$ represents the saturation function of the $i$-th phase with

$$\phi_i(0) = 0, \quad \phi_i(1) = 1, \quad 0 \leq \phi_i(p_i) \leq 1, \quad \frac{d\phi_i}{dp_i} \geq 0 \quad \text{for all} \quad 0 < p_i < 1 . \tag{2.2.40}$$

The simplest case is $\phi(p) := p$. Formula (2.2.39) takes only the growth of a phase fraction into account. This is due to the fact that for the transformation induced plasticity only the currently growing phases $p_i$, thus $\dot{p}_i \geq 0$, are considered.

For further discussion of macroscopic TRIP modelling and for further references we refer to Wolff et al. (2009).

**Remark 2.2.3** (TRIP with back stress). *Generally, TRIP has a back stress $\boldsymbol{X}_{trip}$. Instead of (2.2.39), we have*

$$\dot{\varepsilon}_{trip} = \frac{3}{2} (\boldsymbol{\sigma}^* - \boldsymbol{X}^*_{trip}) \sum_{i=1}^{M_p} \kappa_i \frac{d\phi_i}{dp_i} \max\{\dot{p}_i, 0\} . \tag{2.2.41}$$

*The back stress $\boldsymbol{X}_{trip}$ is defined according to (2.1.18). Using (2.2.37), we obtain*

$$\boldsymbol{X}_{trip} = \varrho_0 \frac{\partial \psi_{in}}{\partial \boldsymbol{\alpha}} = \frac{2}{3} c_{trip}(\theta) \boldsymbol{\alpha} . \tag{2.2.42}$$

*We set the following evolution equation for the internal variable:*

$$\dot{\boldsymbol{\alpha}} = \dot{\varepsilon}_{trip} - \frac{3}{2} \frac{a_{trip}}{c_{trip}} \boldsymbol{X}_{trip} \dot{s}_{trip} , \tag{2.2.43}$$

*where $s_{trip}$ denotes the accumulated TRIP strain (cf. (2.1.54)). Using (2.2.42) and (2.2.43), the evolution of $\boldsymbol{X}_{trip}$ can be described similarly to the evolution of the creep back stress (see (2.2.34)):*

$$\dot{\boldsymbol{X}}_{trip} = \frac{2}{3} c_{trip}(\theta) \dot{\varepsilon}_{trip} - a_{trip}(\theta, \boldsymbol{\sigma}) \boldsymbol{X}_{trip} \dot{s}_{trip} + \frac{\dot{c}_{trip}(\theta)}{c_{trip}(\theta)} \boldsymbol{X}_{trip} , \tag{2.2.44}$$

*where $c_{trip} > 0$, $a_{trip} \geq 0$. See Wolff et al. (2009) for discussion and further references.*

*Here, we do not use a switch $l$ as in (2.2.34). The reason for this is - at least in the isothermal case - that the TRIP back stress should not evolve if the TRIP strain does not evolve (as in classical plasticity).*

The model equations presented above were implemented in order to describe the evolution of creep and TRIP during heating and austenitisation. Chapter 6 will present simulation results as well as experiments investigating the behaviour of the bearing steel SAE 52100 (100Cr6) during heating and austenitisation. A further investigation is presented in Wolff et al. (2012c) and Bökenheide and Wolff (2012).

### 2.2.4 Modelling using multi-mechanism models

The idea of multi-mechanism models (MM models) is the split up of the inelastic strain part of the strain tensor into a sum of several parts, sometimes called *mechanisms*. A model with $m$ mechanisms is referred to as an $m$M model. In this case, the approach for the strain tensor $\varepsilon$ in (2.1.52) is modified to

$$\varepsilon = \varepsilon_{te} + \varepsilon_{in} = \varepsilon_{te} + A_1\varepsilon_1 + A_2\varepsilon_2 + \cdots + A_m\varepsilon_m \,, \qquad (2.2.45)$$

where $A_j$ are positive real numbers ($j = 1 \ldots m$). The elastic strain part is not considered as a mechanism. For an alternative investigation we refer to Kröger (2013) where the elastic strain part is treated as an own mechanism of the model.

A detailed overview of the theory and the application of multi-mechanism models will be presented in Chapter 3. We refer to Wolff and Taleb (2008), Taleb and Cailletaud (2010), Wolff and Böhm (2010), Wolff et al. (2010), Wolff et al. (2011c), Saï (2011), Wolff et al. (2012b) and Kröger (2013) for more information about multi-mechanism models.

In Chapter 3 we will give a general introduction into the topic and present some specific examples of applications. Section 3.3 will present an MM model with two mechanisms for the modelling of creep. A coupled model for creep and TRIP will be presented in Section 3.4.

## 2.3 Modelling of phase transformations during heating and austenitisation

In this Section, we consider a specific type of steel, i.e. SAE 52100 (100Cr6) (cf. results in Chapter 6). In order to describe the occurring phase transformations during heating, we have to consider the evolution of three different phase fractions: the initial material that consists of ferrite and carbide as well as the forming phase, austenite.

We model the transformation of the initial phase into austenite by the dissolution of ferrite and carbide phase fractions. We use an approach of Johnson-Mehl-Avrami-Kolmogoroff (JMAK) type which was presented in Surm et al. (2008). We also refer to Wolff et al. (2012c) and Wolff et al. (2007).

The evolution of the austenite phase fraction $p_A$ depends on the evolution of the carbide fraction $p_C$ in the following way:

$$- \dot{p}_A = k_c \, \dot{p}_C \ . \tag{2.3.1}$$

For unalloyed steels, the parameter $k_c > 0$ is constant. in the case of alloyed steels, $k_c$ has to be determined from experimental data. By (2.1.23) we get

$$\dot{p}_C = \frac{\dot{p}_F}{k_c - 1} \ , \tag{2.3.2}$$

where $p_F$ stands for the ferrite fraction.

We assume that at some time $t = t_A$ ferrite has transformed completely into austenite (where "completely" means that $p_F(x, t_A)$ is smaller than a certain tolerance). Moreover, for $0 \le t < t_A$, ferrite and carbide transform into austenite. Thus, we have the following systems of JMAK type equations, depending on whether $t$ is smaller or larger than $t_A$:

- for $t \in [0, t_A]$:

$$\dot{p}_F(x, t) = -\frac{n_F}{\tau_F(\theta, \nu)} (p_F(x, t) - p_{eq,F}(\theta)) \left( -\ln \left[ \frac{p_F(x, t) - p_{eq,F}(\theta)}{p_{0_F} - p_{eq,F}(\theta)} \right] \right)^{\frac{n_F - 1}{n_F}} ,$$

$$\dot{p}_C(x, t) = \frac{\dot{p}_F(x, t)}{k_c - 1} , \tag{2.3.3}$$

- and for $t > t_A$:

$$\dot{p}_F(x, t) = 0,$$

$$\dot{p}_C(x, t) = -\frac{n_C}{\tau_C(\theta, \nu)} (p_C(x, t) - p_{eq,C}(\theta)) \left( -\ln \left[ \frac{p_C(x, t) - p_{eq,C}(\theta)}{p_{0_C} - p_{eq,C}(\theta)} \right] \right)^{\frac{n_C - 1}{n_C}} . \tag{2.3.4}$$

Here, $\tau_F > 0$, $n_F > 1$, $\tau_C > 0$, $n_C > 1$ are the JMAK parameters for ferrite and carbide, respectively, $\nu$ is the heating rate which is constant in our case, $p_{0,F}$ and $p_{0,C}$ denote the initial ferrite and carbide fractions at $t = 0$ and $p_{eq,F}(\theta)$, $p_{eq,C}(\theta)$ stand for the equilibrium phase fractions at temperature $\theta$. The parameters $\tau_F$, $\tau_C$ are modelled by

$$\tau_i(\theta, \nu) = \tau_{0,i}(\nu) \exp \left( \frac{-Q}{R \, \theta_{abs}} \right) \tag{2.3.5}$$

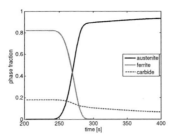

Figure 2.1: Phase fractions during heating.

where $i$ stands for $F$ and $C$, respectively, $Q$ denotes the activation energy for carbon diffusion in austenite, $R$ stands for the gas constant and $\theta_{abs}$ denotes the absolute temperature in Kelvin (cf. Remark 2.3.1).

By means of the single phase fractions $p_F$, $p_C$ we obtain the bulk density $\varrho(\theta, p)$ at current temperature $\theta$ and phase fractions $p$ by the mixture rule

$$\varrho(\theta, p) = \varrho_F(\theta) \, p_F + \varrho_C(\theta) \, p_C + \varrho_A(\theta, u_c) \, p_A \,, \tag{2.3.6}$$

where $\varrho_i$ stands for the density of the respective phase $p_i$. The density of austenite additionally depends on the carbon content in austenite (see Surm et al. (2008) and the references therein). The vector $p$ consists of the single phase fractions $p_i$. The single phase fractions $p_i$ are summarised in the vector $p$ (cf. Section 2.1.4).

After having determined the single phase fractions as well as the creep strains of the different phases, one obtains the total creep strain of the phase mixture by the mixture rule

$$\varepsilon_c = p_F \varepsilon_{c,F} + p_C \, \varepsilon_{c,C} + p_A \, \varepsilon_{c,A}, \tag{2.3.7}$$

where $\varepsilon_{c,i}$ stand for the creep strains of the respective phases $p_i$, each represented by an evolution equation (see (2.2.30)). The corresponding material parameters in the evolution equations referring to the single creep strains of the different phases are determined beforehand using data from isothermal experiments with the initial material as well as with austenite. The results of these parameter identifications will be presented in Chapter 6.

**Remark 2.3.1.**    *(i) We consider the transformation of the initial material, which is in our case annealing on globular cementite, into austenite. Here, the transformation of the initial material is modelled by the dissolution of the two phases ferrite and carbide. Besides this, it is also possible to consider the initial material as one phase and to model the transformation of this phase into the second phase austenite.*
*But, further results have shown that this approach is too imprecise. Therefore, we chose the approach presented above.*

*(ii) For the activation energy $Q$ in (6.2.2) we use $Q = 141\,kJ/mol$. The gas constant is $R = 8.314\,J/(K\,mol)$.*

## 2.4 Problem setting

Summarising the material laws presented in the last sections, we are able to describe the material behaviour of steel during heating and austenitisation. The model equations are presented in Box 2.4.1 Altogether, we obtain a coupled non-linear initial-boundary value problem of ordinary and partial differential equations.

Chapters 4 and 5 will cover the discretisation of the presented model. The discretised equations were implemented in order to perform simulations in a one-dimensional as well as in a three-dimensional background. The results of these simulations as well as the validation with experimental data will be presented in Chapter 6.

### 2.4.1. Coupled problem

- **Heat equation**

$$\varrho_0 c_d(\theta,p)\frac{\partial\theta}{\partial t} - \operatorname{div}\left(k(\theta,p)\nabla\theta\right) = \varrho_0\sum_{i=2}^{M_p} L_i(\theta)\dot p_i \qquad \text{in}\quad \Omega\times(0,T)$$

$$\theta(x,0)=\theta_0 \qquad\qquad\qquad \text{in}\quad \Omega$$

$$\theta(x,t)=\theta_{ext_D}(t) \qquad\qquad \text{on}\quad \partial\Omega\times(0,T_1)$$

$$-k_\theta(\theta,p)\nabla\theta\cdot n = \delta(\theta(x,t)-\theta_{ext_R}(t)) \quad \text{on}\quad \partial\Omega\times(T_1,T)$$

- **Phase transformations**

For $t\in[0,t_A]$:

$$\dot p_F(x,t) = -\frac{n_F}{\tau_F(\theta,\nu)}(p_F(x,t)-p_{eq,F}(\theta))\left(-\ln\left[\frac{p_F(x,t)-p_{eq,F}(\theta)}{p_{0_F}-p_{eq,F}(\theta)}\right]\right)^{\frac{n_F-1}{n_F}}$$

$$\dot p_C(x,t) = \frac{\dot p_F(x,t)}{k_c-1}$$

For $t>t_A$:

$$\dot p_F(x,t) = 0$$

$$\dot p_C(x,t) = -\frac{n_C}{\tau_C(\theta,\nu)}(p_C(x,t)-p_{eq,C}(\theta))\left(-\ln\left[\frac{p_C(x,t)-p_{eq,C}(\theta)}{p_{0_C}-p_{eq,C}(\theta)}\right]\right)^{\frac{n_C-1}{n_C}}$$

where $\tau_F>0$, $n_F>1$, $\tau_C>0$, $n_C>1$

- **Mechanics**

  - Deformation equation

$$\varrho_0\,\ddot u - \operatorname{div}\sigma = \varrho_0 f \qquad \text{in}\quad \Omega\times(0,T)$$

$$u(x,0)=0\quad,\quad \dot u(x,0)=0, \qquad \text{in}\quad \Omega$$

$$u(x,t)=g_D \qquad\qquad \text{on}\quad \Gamma_D\times(0,T)$$

$$\sigma\cdot n = g_N \qquad\qquad \text{on}\quad \Gamma_N\times(0,T)$$

  - Strain and stress tensors

$$\varepsilon(u)=\frac{1}{2}\left(\nabla u+(\nabla u)^T\right)$$

$$\varepsilon = \varepsilon_{te}+\varepsilon_{in}=\varepsilon_{te}+\varepsilon_c+\varepsilon_{trip}$$

$$\sigma = 2\mu(\theta,p)(\varepsilon-\varepsilon_{in})+\lambda(\theta,p)(\operatorname{tr}(\varepsilon))I - 3K(\theta,p)\left(\sqrt[3]{\frac{\varrho_0}{\varrho(\theta,p)}}-1\right)I$$

- Creep strain

$$\dot{\boldsymbol{\varepsilon}}_c = \frac{3}{2} A(\theta) \left( \sqrt{\frac{3}{2}} \frac{\|\boldsymbol{\sigma}^* - \boldsymbol{X}_c^*\|}{D_c} \right)^{m(\theta)-1} \frac{\boldsymbol{\sigma}^* - \boldsymbol{X}_c^*}{D_c} s_c^{k(\theta)}$$

$$\dot{\boldsymbol{X}}_c = \frac{2}{3} c_c(\theta) \dot{\boldsymbol{\varepsilon}}_c - b_c(\theta, \boldsymbol{\sigma}) \boldsymbol{X}_c (\dot{s}_c)^l + \frac{\mathrm{d}c_c}{\mathrm{d}\theta} \frac{\dot{\theta}}{c_c(\theta)} \boldsymbol{X}_c$$

with $A > 0$, $m > 0$, $c_c > 0$, $b_c \geq 0$, $l \in \{0, 1\}$

- TRIP strain

$$\dot{\boldsymbol{\varepsilon}}_{trip} = \frac{3}{2} \boldsymbol{\sigma}^* \sum_{i=1}^{M_p} \kappa_i \frac{\mathrm{d}\phi_i}{\mathrm{d}p_i} \max\{ \dot{p}_i, 0 \}$$

with $\kappa_i > 0$, $\phi_i(0) = 0$, $\phi_i(1) = 1$, $0 \leq \phi_i(p_i) \leq 1$, $\frac{\mathrm{d}\phi_i}{\mathrm{d}p_i} \geq 0$ for all $0 < p_i < 1$.

# 3 Multi-mechanism models

This Chapter handles a special modelling approach: the theory of multi-mechanism models. The first Section will give a general introduction into the topic. Generally, multi-mechanism models are applied to model ratcheting (see Saï and Cailletaud (2007), e.g.) in metal plasticity and the material behaviour of steel during phase transformations. This modelling approach represents an extension of the classical material law for inelastic material behaviour by decomposing the inelastic strain into several parts. These parts are sometimes called *mechanisms*. In contrast to rheological models, there can be an interaction between the single mechanisms. The idea of multi-mechanism models is to obtain further possibilities in the modelling of material behaviour. For instance, in the case of classical plasticity, each mechanism can possess its own yield stress. The special feature of this modelling approach lies in the fact that the mechanisms can interact with each other.

This chapter is structured as follows: First, we will give a general introduction and provide the basic equations. After that, we will focus especially on two-mechanism models (abbreviated as 2M models). Section 3.2 will present the model equations of two different types of 2M models. In Section 3.3 we will apply the presented model approach to our problem presented in Chapter 2. We will present two types of 2M models in order to model creep material behaviour of steel. After that, in Section 3.4, we will present an example of a 2M model that is used to model creep and TRIP arising simultaneously. Here, the two inelastic strains are independent from each other but coupled via their back stresses.

Furthermore, we will verify the thermodynamic consistency of the presented models (see Sections 3.3.3 and 3.4.2).

## 3.1 General introduction

For a general introduction into the theory of multi-mechanism models, we refer to Besson et al. (2001), Wolff and Taleb (2008), Taleb and Cailletaud (2010), Wolff et al. (2010) and Wolff et al. (2011c). Multi-mechanism models for the description of ratcheting are investigated in Saï and Cailletaud (2007). An overview of different multi-mechanism models can be found in Saï (2011). Here, mainly plastic mechanisms are considered.

Besides classical plasticity, there are further investigations covering different material behaviour and the modelling via mutli-mechanism models: In Kröger (2013), the author presents the application for different material behaviour. Especially viscoelasticity in the framework of multi-mechanism models is investigated. Another approach for thermo-viscoelasticity via a two-mechanism model can be found in

Wolff et al. (2012b). In Wolff and Böhm (2010), a 2M model with creep mechanisms is presented. Viscoelastic and creep material behaviour are considered as phenomena with a yield stress equal to zero (cf. Section 3.3). A study of plastic and viscoplastic models can be found in Cailletaud and Saï (1995). In Wolff et al. (2013) and Wolff et al. (2015), the authors present an extended generalised approach of multi-mechanism models used for the modelling of classical plastic material behaviour.

In the following, we will first provide the continuum-mechanical preparations for inelastic material behaviour in the case of small deformations. Section 2.1 presented the required model equations.

### 3.1.1 General multi-mechanism models (in series)

Here, we focus on multi-mechanism models in serial connection[1]. We use definition (2.1.4) of the total strain tensor $\varepsilon$. The idea of multi-mechanism models is a further splitting of the inelastic strain part $\varepsilon_{in}$ of the total strain (see (2.1.5)).

Generally, the inelastic strain is split into a sum of $m$ parts (or 'mechanisms'):

$$\varepsilon_{in} = A_1\varepsilon_1 + A_2\varepsilon_2 + \ldots + A_m\varepsilon_m \,, \qquad (3.1.1)$$

where $A_j \in \mathbb{R}$, $A_j > 0$ for $j = 1, \ldots, m$. This is referred to as an $m$-mechanism model, abbreviated as $m$M model. The parameters $A_1, \ldots, A_m$ offer opportunities for further extensions, see Remark 3.1.1 for details. We also refer to Wolff et al. (2011c) and Saï and Cailletaud (2007), e.g.

The partial inelastic strains $\varepsilon_j$ are assumed to be traceless, i.e.

$$\mathrm{tr}(\varepsilon_1) = \mathrm{tr}(\varepsilon_2) = \ldots = \mathrm{tr}(\varepsilon_m) = 0 \,. \qquad (3.1.2)$$

Usually, the elastic strain part $\varepsilon_{te}$ is not regarded as a mechanism. An alternative approach is presented in Kröger (2013) where the elastic strain is treated as a mechanism.

The theory of multi-mechanism models represents a generalisation of the classical approach based on rheological models. For a general introduction to rheological models, we refer to Altenbach and Altenbach (1994), Palmov (1998) and Besson et al. (2001), e.g..

The important fact considering $m$M models - in contrast to rheological models - is that there is an *interaction* between the single mechanisms. Figure 3.1 illustrates a two-mechanism model. The advantage of $m$M models consists in the larger number of possibilities in modelling the material behaviour. For an extended and more general approach, see Wolff et al. (2013) and Wolff et al. (2015).

For each partial strain $\varepsilon_j$ we introduce a corresponding accumulated strain $s_j$ by

$$s_j(t) := \int_0^t \sqrt{\frac{2}{3}\dot{\varepsilon}_j(\tau) : \dot{\varepsilon}_j(\tau)} \, \mathrm{d}\tau \quad , \quad j = 1, \ldots, m \,. \qquad (3.1.3)$$

---

[1]Some remarks about multi-mechanism models in parallel connection are given in Wolff et al. (2015).

Note that the accumulation of the total inelastic strain $\boldsymbol{\varepsilon}_{in}$ is *not* represented by the sum of the single accumulations $s_j$.

Furthermore, we introduce the *local* or *partial stresses* $\boldsymbol{\sigma}_j$ by

$$\boldsymbol{\sigma}_j := A_j \boldsymbol{\sigma} \quad , \quad j = 1, \ldots, m \; . \tag{3.1.4}$$

where $A_j$ are positive real numbers (cf. Remark 3.1.1 for extensions and Wolff et al. (2015) for further explanations). Using the split of the strain tensor $\boldsymbol{\varepsilon}_{in}$ in (3.1.1) and the definition of the local stresses (3.1.4), we have

$$\boldsymbol{\sigma} : \boldsymbol{\varepsilon}_{in} = \boldsymbol{\sigma} : \left( \sum_{j=1}^{m} A_j \boldsymbol{\varepsilon}_j \right) = \sum_{j=1}^{m} (\boldsymbol{\sigma}_j : \boldsymbol{\varepsilon}_j) \, , \tag{3.1.5}$$

and the remaining inequality (2.1.19) takes the form

$$\sum_{j=1}^{m} \boldsymbol{\sigma}_j : \dot{\boldsymbol{\varepsilon}}_j - \sum_{j=1}^{N_\xi} \boldsymbol{\chi}_j : \dot{\boldsymbol{\xi}}_j \geq 0 \; . \tag{3.1.6}$$

Each partial strain $\boldsymbol{\varepsilon}_j$ has its own evolution equation which will be derived from the specific material behaviour (e.g. plastic or creep mechanisms). The mechanisms are not independent from each other, each mechanism can possess an own criterion or there can be several common criteria.

We will introduce the corresponding criteria of the mechanisms in the following, depending on the material behaviour. For instance, in the case of a plastic mechanism, the criterion is represented by the flow function.

A model with $m$ mechanisms and $n$ criteria is referred to as an $m$M$n$C model in the following, where always $n \leq m$. In Wolff et al. (2013) and Wolff et al. (2015), the authors represent a more general approach of $m$M$n$C models. For a general overview of different multi-mechanism models we refer to Saï (2011). Different examples of mechanisms and models are given in Kröger (2013).

If the mechanisms model *different* material behaviour the mechanisms might be modelled independently from each other. However, they can still be coupled via their back stresses. An example for this will be given in Section 3.4 where a 2M model is presented consisting of creep and TRIP mechanisms.

**Remark 3.1.1** (Weighting parameters).
*In 3.1.1, we assume the weighting parameters $A_j$ for $j = 1, \ldots, m$ to be positive real numbers. In most contributions, these parameters are fixed to one. Examples for $A_j \neq 1$ are given in Saï (1993).*

*Furthermore, the approach can be extended by considering $A_j$ as (time- and space-dependent) functions. They can constitute for instance phase fractions in complex materials. This approach can capture effects at the micro scale. We refer to Saï and Cailletaud (2007) and Saï et al. (2011). We refer to Wolff et al. (2015) for explanations and further references. In this case, there arise additional terms in the remaining inequality (3.1.6). In order to ensure thermodynamic consistency, one has to assume suitable evolution equations, see Mahnken et al. (2015).*

*More complex multi-mechanism models can be found in* Regrain et al. (2009). *Instead of* (3.1.4), *the partial stresses are defined with further variables.*

In the following Sections, we will especially focus on two-mechanism models. We will deal with 2M models with either one common criterion or two criteria, abbreviated as 2M1C and 2M2C models, respectively. We will examplarily introduce 2M models for classic plasticity first (see Section 3.2).

After this, we will investigate the modelling of creep by means of two-mechanism models (see Section 3.3). We will consider our model presented in Chapter 2, see Section 2.4. Here, we exploit the fact that one is able to model creep similarly to the modelling of plasticity. We will set the yield stress formally equal to zero and we will state a yield function such that the mechanism is always active. For details we refer to Section 3.3.

In Section 3.4, another example of a two-mechanism model is presented where the mechanisms model creep and TRIP arising simultaneously.

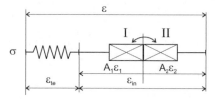

Figure 3.1: Scheme of multi-mechanism model consisting of a thermo-elastic element and two inelastic elements (from Wolff et al. (2015)).

## 3.2 Two-mechanism models

In the following, we focus on multi-mechanism models with two mechanisms, abbreviated as 2M models. We will consider two specific examples: a 2M model with one criterion as well as with two criteria.

As multi-mechanism models were originally developed in order to describe especially (classical) plasticity or viscoplasticity, we will exemplarily introduce a 2M model to describe classic plastic material behaviour first. In the following Sections 3.2.1 and 3.2.2 we will present the model equations for plastic mechanisms. After that, Sections 3.3 and 3.4, will deal with a two-mechanism model for creep as well as for creep and TRIP, respectively.

There are many publications dealing with MM models describing metal plasticity or ratcheting. We refer to Besson et al. (2001), Cailletaud and Saï (1995), Wolff and Taleb (2008), Taleb and Cailletaud (2010), Wolff et al. (2010) Wolff et al. (2011c), Saï and Cailletaud (2007) and Saï (2011), e.g. For further investigations on two-mechanism models we refer to Wolff et al. (2010),Wolff et al. (2011c) and Wolff et al. (2012b), e.g. In Wolff and Böhm (2010), the authors focus on the modelling of creep via two-mechanism models.

Considering the split of the inelastic strain in (3.1.1), we set $m = 2$ in the following. Thus, the inelastic strain is given by

$$\varepsilon_{in} = A_1 \varepsilon_1 + A_2 \varepsilon_2 \quad , \quad \text{where} \quad A_1, A_2 > 0 \quad \text{and} \quad \text{tr}(\varepsilon_1) = \text{tr}(\varepsilon_2) = 0 \ . \quad (3.2.1)$$

Here, the parameters $A_1, A_2$ are real numbers (cf. Remark 3.1.1 for extensions). The local stresses (cf. equation (3.1.4)) are defined by

$$\sigma_j := A_j \sigma \quad , \quad j = 1, 2 \ . \quad (3.2.2)$$

In the next step, we will specialise the ansatz for the inelastic part of the free energy in (2.1.9). First, we introduce the split

$$\psi_{in} = \psi_{kin} + \psi_{iso} \,, \quad (3.2.3)$$

such that the inelastic part of the free energy is decomposed into two parts corresponding to kinematic and isotropic hardening, respectively. Hence, here we do not consider a coupling between these two parts (cf. Remark 3.2.1).

The definition of $\psi_{in}$ depends on the particular chosen model, i.e. on the number of mechanisms and criteria of the model. First, we will consider a 2M model with one common criterion. After that, we focus on a 2M2C model which contains one criterion for each mechanism.

In this section, we will focus on plastic mechanisms taking kinematic as well as isotropic hardening into account (cf. Section 2.2.1).

**Remark 3.2.1.** *(i) Here, we do not consider a coupling between kinematic and isotropic hardening. We refer to* Cailletaud and Saï (1995) *and* Wolff et al. (2010) *for this extension.*

*(ii) Considering multi-mechanism models, we focus especially on models with two mechanisms. The transition to m is without difficulty.*

### 3.2.1 Two-mechanism models with one criterion (2M1C) for plasticity

We specialise the ansatz for the inelastic part of the free energy $\psi_{in}$ in (2.1.7) taking into account the split (3.2.3). We assume the internal variables to be given as $\xi = (\alpha_1, \alpha_2, r)$, thus

$$\psi_{in} = \psi_{in}(\alpha_1, \alpha_2, r, \theta) = \psi_{kin}(\alpha_1, \alpha_2, \theta) + \psi_{iso}(r, \theta) \ . \quad (3.2.4)$$

Here, the internal variables $\xi$ consist of two symmetric tensorial variables of strain type $\alpha_j$ - one for each mechanism $\varepsilon_j$, $j = 1, 2$, respectively - and one scalar internal variable $r$. Considering classical plasticity, the tensorial internal variables are related to kinematic hardening, the scalar internal variable to isotropic hardening.

**Remark 3.2.2** (Internal variables). *In the case of creep, we only consider kinematic hardening. We will set $\xi = (\alpha_1, \alpha_2)$ for the internal variables of a 2M model, see Section 3.3 for details.*

First, we specify the inelastic part of the free energy (2.1.9) by defining the single parts in (3.2.4). We suppose the following ansatz for a 2M1C model:

$$\psi_{kin}(\boldsymbol{\alpha}_1, \boldsymbol{\alpha}_2, \theta) := \frac{1}{3\varrho_0}\left(c_{11}(\theta)\boldsymbol{\alpha}_1 : \boldsymbol{\alpha}_1 + 2c_{12}(\theta)\boldsymbol{\alpha}_1 : \boldsymbol{\alpha}_2 + c_{22}(\theta)\boldsymbol{\alpha}_2 : \boldsymbol{\alpha}_2\right),$$

$$(3.2.5)$$

and

$$\psi_{iso}(r, \theta) := \frac{1}{2\varrho_0}d(\theta)r^2 . \qquad (3.2.6)$$

Assuming

$$c_{11}(\theta) > 0, \quad c_{12}^2(\theta) \le c_{11}(\theta)c_{22}(\theta), \quad \text{and} \quad d(\theta) \ge 0, \qquad (3.2.7)$$

for all admissible temperatures $\theta$, the inelastic free energy $\psi_{in}$ is a convex and non-negative function.

Next, we define the back stresses as thermodynamic forces (see (2.1.18)) via the partial derivative of the free energy with respect to the internal variables related to kinematic hardening:

$$\boldsymbol{X}_j = \varrho_0\frac{\partial\psi_{kin}}{\partial\boldsymbol{\alpha}_j}, \quad j = 1, 2 . \qquad (3.2.8)$$

Taking the ansatz (3.2.5) for the kinematic part of $\psi_{in}$ into account, we obtain:

$$\boldsymbol{X}_1 = \frac{2}{3}c_{11}\boldsymbol{\alpha}_1 + \frac{2}{3}c_{12}\boldsymbol{\alpha}_2 \qquad (3.2.9a)$$

$$\boldsymbol{X}_2 = \frac{2}{3}c_{12}\boldsymbol{\alpha}_1 + \frac{2}{3}c_{22}\boldsymbol{\alpha}_2 . \qquad (3.2.9b)$$

In the same way, we define the isotropic hardening stress $R$ as thermodynamic force using (2.1.18) and (3.2.6):

$$R = \varrho_0\frac{\partial\psi_{iso}}{\partial r} = d(\theta)r . \qquad (3.2.10)$$

By means of the introduced definitions of the inelastic part of the free energy $\psi_{in}$ and of the internal variables $\xi$, the remaining inequality (2.1.19) can be specified. We use (3.2.5),(3.2.6) and the equations for the partial strains and stresses (3.2.1), (3.1.4), respectively, as well as the definitions (3.2.9) and (3.2.10).

The remaining inequality (3.1.6) reads as

$$\boldsymbol{\sigma}_1 : \dot{\boldsymbol{\varepsilon}}_1 + \boldsymbol{\sigma}_2 : \dot{\boldsymbol{\varepsilon}}_2 - \boldsymbol{X}_1 : \dot{\boldsymbol{\alpha}}_1 - \boldsymbol{X}_2 : \dot{\boldsymbol{\alpha}}_2 - R\dot{r} \ge 0 . \qquad (3.2.11)$$

This expression can be rewritten in the form

$$(\boldsymbol{\sigma}_1 - \boldsymbol{X}_1) : \dot{\boldsymbol{\varepsilon}}_1 + (\boldsymbol{\sigma}_2 - \boldsymbol{X}_2) : \dot{\boldsymbol{\varepsilon}}_2 + \boldsymbol{X}_1 : (\dot{\boldsymbol{\varepsilon}}_1 - \dot{\boldsymbol{\alpha}}_1) + \boldsymbol{X}_2 : (\dot{\boldsymbol{\varepsilon}}_2 - \dot{\boldsymbol{\alpha}}_2) - R\dot{r} \ge 0 .$$

$$(3.2.12)$$

In order to study thermodynamical consistency, this form of the remaining inequality in (3.2.12) is more convenient (cf. Wolff and Taleb (2008), Wolff et al. (2011c), e.g.). In the following subsection, we give some further details.

Based on the von Mises stress, we define the expressions

$$J_j := \sqrt{\frac{3}{2}} \| \boldsymbol{\sigma}_j^* - \boldsymbol{X}_j^* \| = \sqrt{\frac{3}{2} (\boldsymbol{\sigma}_j^* - \boldsymbol{X}_j^*) : (\boldsymbol{\sigma}_j^* - \boldsymbol{X}_j^*)} \, , \quad j = 1, 2 \, , \tag{3.2.13}$$

and

$$J := \left( J_1^N + J_2^N \right)^{\frac{1}{N}} \, , \quad \text{with} \quad N > 1 \, . \tag{3.2.14}$$

Remarks on the importance of the parameter $N$ in (3.2.14) can be found in Wolff et al. (2010),Wolff and Taleb (2008) and Taleb and Cailletaud (2010).

Considering two-mechanism models with one criterion, we define a *common* yield function $f$ by

$$f(\boldsymbol{\sigma}_1, \boldsymbol{\sigma}_2, \boldsymbol{X}_1, \boldsymbol{X}_2, R, R_0) := J - (R + R_0) \, , \tag{3.2.15}$$

where $R_0 := \sqrt[N]{2} \, \sigma_0$ and $\sigma_0$ denotes the initial yield stress. In the case of plastic mechanisms, this function distinguishes the elastic from the plastic domain. For a 2M1C model with plastic mechanisms one supposes the constraint

$$f(\boldsymbol{\sigma}_1, \boldsymbol{\sigma}_2, \boldsymbol{X}_1, \boldsymbol{X}_2, R, R_0) \leq 0 \, . \tag{3.2.16}$$

We refer to Wolff et al. (2010) and Wolff et al. (2011c) for a detailed description of 2M models with plastic mechanisms.

Next, our aim is to formulate evolution equations for the inelastic strains. Based on (3.2.13), (3.2.14) and (3.2.15), we define

$$n_j := -\frac{\partial f}{\partial \boldsymbol{X}_j} = \frac{3}{2} \frac{\boldsymbol{\sigma}_j^* - \boldsymbol{X}_j^*}{J_j} \left( \frac{J_j}{J} \right)^{N-1} \, , \quad j = 1, 2 \, . \tag{3.2.17}$$

Generally, all mechanisms belonging to the same flow criterion have a common plastic multiplier. Therefore, we assume the following evolution laws for the two inelastic mechanisms $\boldsymbol{\varepsilon}_1, \boldsymbol{\varepsilon}_2$:

$$\dot{\boldsymbol{\varepsilon}}_j = \gamma \, n_j \, , \quad j = 1, 2 \, , \tag{3.2.18}$$

where $\gamma$ denotes the common plastic multiplier for both mechanisms.

The plastic multiplier has to fulfil:

$$\gamma = 0 \quad \text{if} \quad f(\boldsymbol{\sigma}_1, \boldsymbol{\sigma}_2, \boldsymbol{X}_1, \boldsymbol{X}_2, R, R_0) < 0 \, , \tag{3.2.19a}$$
$$\gamma \geq 0 \quad \text{if} \quad f(\boldsymbol{\sigma}_1, \boldsymbol{\sigma}_2, \boldsymbol{X}_1, \boldsymbol{X}_2, R, R_0) = 0 \, . \tag{3.2.19b}$$

The following Section will handle creep mechanisms. In this case, the multiplier is not determined via a consistency condition as in (3.2.19) but can be stated directly (cf. (3.3.16) and (3.3.32)). See Section 3.3 for details.

For the partial accumulated strains defined in (3.1.3), we obtain by means of the definitions (3.2.13),(3.2.14),(3.2.17) and the evolution law (3.2.18):

$$\dot{s}_j = \gamma \left( J_1^N + J_2^N \right)^{\frac{1-N}{N}} J_j^{N-1} \, , \quad j = 1, 2 \, . \tag{3.2.20}$$

Then, for the plastic multiplier $\gamma$ follows that

$$\gamma = \left( \dot{s}_1^{\frac{N}{N-1}} + \dot{s}_2^{\frac{N}{N-1}} \right)^{\frac{N-1}{N}} . \tag{3.2.21}$$

The inelastic multiplier $\gamma$ is a common multiplier for both mechanisms. Furthermore, we assume an evolution equation for the internal variable $r$:

$$\dot{r} = \gamma - aR\gamma , \tag{3.2.22}$$

with $a > 0$. Finally, there remains an approach for the evolution equations describing $\boldsymbol{\alpha}_1$ and $\boldsymbol{\alpha}_2$. We set:

$$\dot{\boldsymbol{\alpha}}_j = \dot{\boldsymbol{\varepsilon}}_j - \frac{3}{2} \sum_{i=1}^{2} b_{ji} \boldsymbol{X}_i \gamma \quad , \quad j = 1, 2 , \tag{3.2.23}$$

where the matrix $\boldsymbol{b}$ containing the coefficients $b_{ji}$ is assumed to be positive semi-definite. Using (3.2.9) with (3.2.23) yields the generalised Armstrong-Frederick equations for the evolution of the back stresses $\boldsymbol{X}_1, \boldsymbol{X}_2$.

In Section 3.3.1, we will present a 2M1C model that is used to model creep. The model presented above is simplified and we will introduce approaches for multiplier and internal variables.

**Remark 3.2.3** (Viscoplastic mechanisms).
*In the same way as presented above in Section 3.2.1 for plastic mechanisms (and in 3.3.1 for creep mechanisms), viscoplastic mechanisms can be handled. But, in contrast to classical plasticity, the viscoplastic multiplier is not determined via a consistency condition, but will be stated directly. Both material behaviours - plastic as well as viscoplastic - are characterised by the division into an elastic and an inelastic domain.*

*In the case of a 2M1C model, we define the yield function $f_{vp}$ as in (3.2.15) and $n_1, n_2$ according to (3.2.17). Formally, the evolution laws of $\boldsymbol{\varepsilon}_{vp1}, \boldsymbol{\varepsilon}_{vp2}$ are as in (3.2.18). We have:*

$$\dot{\boldsymbol{\varepsilon}}_{vpj} = \gamma_{vp} n_j \quad for \quad j = 1, 2 . \tag{3.2.24}$$

*The elastic domain is defined by*

$$f_{vp}(\boldsymbol{\sigma}_1, \boldsymbol{\sigma}_2, \boldsymbol{X}_1, \boldsymbol{X}_2, R, R_0) \leq 0 . \tag{3.2.25}$$

*The viscoplastic multiplier is not determined by flow and consistency conditions but must be defined separately. We set (following the approach in* Wolff et al. (2011c)):

$$\gamma_{vp} := \frac{2}{3\eta} \left\langle \frac{1}{D} f_{vp}(\boldsymbol{\sigma}_1, \boldsymbol{\sigma}_2, \boldsymbol{X}_1, \boldsymbol{X}_2, R, R_0) \right\rangle^n , \tag{3.2.26}$$

*where $n > 0$, $\eta > 0$ stands for the viscosity (generally depending on temperature and maybe on other quantities) and $D$ is a positive scalar denoting the drag stress* *(see* Chaboche (2008) *for details). In* (3.2.26), $< \cdot >$ *denote the Mc Cauley brackets which are defined by*

$$< x > := \begin{cases} x & for \ x \geq 0 \\ 0 & for \ x < 0 \end{cases} . \tag{3.2.27}$$

*Thus, in contrast to classical plasticity, we have* $\gamma \geq 0$ *if* $f_{vp} > 0$. *Furthermore, the relations* (3.2.21) *and* (3.2.20), *are also valid for* $\gamma_{vp}$ *and* $s_{vp1}, s_{vp2}$.

*We refer to* Chaboche (2008), Haupt (2002), de Souza Neto et al. (2008), *e.g. for a general introduction to viscoplasticity.*

| | **Plasticity** | **Viscoplasticity** | **Creep** |
|---|---|---|---|
| Kinematic hardening | + | + | + |
| Isotropic hardening | + | + | $-^2$ |
| Multiplier | via consistency condition | $\gamma_{vp} =< \frac{f}{D} >$ | via material law |
| Yield criterion/ criteria | $f = 0$ $f_i = 0$ | $f > 0$ $f_i > 0$ | ✓ ✓ |

Table 3.1: Comparison between the modelling possibilities of plastic, viscoplastic and creep material behaviour. In the case of creep, $\gamma_c > 0$ is always fulfilled.

**Thermodynamic consistency**

In order to prove the thermodynamic consistency of the presented 2M1C model, we have to verify the remaining inequality (3.2.12). Therefore, we insert (3.2.18), (3.2.23) and (3.2.22) in (3.2.12) and prove the validity of:

$$\gamma(\boldsymbol{\sigma}_1 - \boldsymbol{X}_1) : n_1 + \gamma(\boldsymbol{\sigma}_2 - \boldsymbol{X}_2) : n_2 +$$
$$+ \frac{3}{2} \gamma \, \boldsymbol{X}_1 : \left( \sum_{i=1}^{2} b_{1i} \boldsymbol{X}_i \right) + \frac{3}{2} \gamma \, \boldsymbol{X}_2 : \left( \sum_{i=1}^{2} b_{2i} \boldsymbol{X}_i \right) - R\gamma \, (1 \ - aR) \ \geq \ 0 , \tag{3.2.28}$$

which can be rewritten into

$$\gamma \left( R_0 + aR^2 \right) + \frac{3}{2} \gamma \sum_{i,j=1}^{2} b_{ij} \boldsymbol{X}_i : \boldsymbol{X}_j \ \geq 0 , \tag{3.2.29}$$

where we use definition (3.2.17) and take into account that

$$\sum_{j=1}^{2} \gamma(\boldsymbol{\sigma}_j - \boldsymbol{X}_j) : n_j = \gamma J \quad \text{and} \quad J = R_0 + R . \tag{3.2.30}$$

---

[2] Thus, $R \equiv 0$, $R_0 = 0$ in the case of creep.

The last equality follows from (3.2.15) and the yield criterion (3.2.19b), i.e. $f = 0$ in case of plasticity (otherwise $\gamma = 0$). Due to the non-negativity of $\gamma$ and as $a > 0$, the inequality (3.2.29) is fulfilled under the assumption that the matrix $\boldsymbol{b}$ consisting of the coefficients $b_{ij}$ is positive semi-definite.

For further details we refer to Wolff and Taleb (2008). In Wolff et al. (2010) and Wolff et al. (2011c) the authors present an extended version of the evolution equations. The proof of thermodynamic consistency requires therefore additional conditions.

In the following Sections we will present some examples of 2M models and prove their thermodynamic consistency. Section 3.3 will present some 2M models with creep mechanisms, Section 3.4 handles a model with creep and TRIP mechanisms. See 3.3.3 and 3.4.2 for details about the thermodynamic consistency of the presented models.

**Remark 3.2.4** (Thermodynamic consistency in the viscoplastic case).
*In contrast to classical plasticity, for the viscoplastic multiplier $\gamma_{vp}$ (see (3.2.26)) holds that*

$$\gamma_{vp} > 0 \quad if \quad J > R_0 + R \tag{3.2.31}$$

*whereas the plastic multiplier $\gamma$ is only positive if $J = R_0 + R$. Therefore, (3.2.29) is also sufficient in order to prove the thermodynamic consistency in the viscoplastic case.*

## 3.2.2 Two-mechanism models with two criteria (2M2C) for plasticity

We specialise the ansatz for the inelastic part of the free energy $\psi_{in}$ (see (2.1.7) and (3.2.3)) for the case of a 2M model with two criteria. In the case of a model with two criteria, we assign one internal variable related to isotropic hardening to each flow criterion. Therefore, we assume the internal variables to be given as $\xi = (\boldsymbol{\alpha}_1, \boldsymbol{\alpha}_2, r_1, r_2)$. We have

$$\psi_{in} = \psi_{in}(\boldsymbol{\alpha}_1, \boldsymbol{\alpha}_2, r_1, r_2, \theta) = \psi_{kin}(\boldsymbol{\alpha}_1, \boldsymbol{\alpha}_2, \theta) + \psi_{iso}(r_1, r_2, \theta) . \tag{3.2.32}$$

Here, the internal variables $\xi$ consist of two symmetric tensorial variables of strain type $\boldsymbol{\alpha}_j$ as well as two scalar internal variables $r_j$ - one for each mechanism - where $j = 1, 2$, respectively. The tensorial internal variables $\boldsymbol{\alpha}_1, \boldsymbol{\alpha}_2$ are related to kinematic hardening, the scalar internal variables $r_1, r_2$ to isotropic hardening.

We specify the single parts of $\psi_{in}$ in (3.2.32):

$$\psi_{kin}(\boldsymbol{\alpha}_1, \boldsymbol{\alpha}_2, \theta) := \frac{1}{3\varrho_0}\left(c_{11}(\theta)\boldsymbol{\alpha}_1 : \boldsymbol{\alpha}_1 + 2c_{12}(\theta)\boldsymbol{\alpha}_1 : \boldsymbol{\alpha}_2 + c_{22}(\theta)\boldsymbol{\alpha}_2 : \boldsymbol{\alpha}_2\right),$$
(3.2.33)

and

$$\psi_{iso}(r_1, r_2, \theta) := \frac{1}{2\varrho_0}\left(d_{11}(\theta)r_1^2 + 2d_{12}(\theta)r_1r_2 + d_{22}(\theta)r_2^2\right).$$
(3.2.34)

The inelastic free energy $\psi_{in}$ is convex, if

$$c_{11}(\theta) > 0, \quad c_{12}^2(\theta) \leq c_{11}(\theta)c_{22}(\theta), \quad d_{11}(\theta) \geq 0 \quad \text{and} \quad d_{12}^2(\theta) \leq d_{11}(\theta)d_{22}(\theta),$$
(3.2.35)

for all admissible temperatures $\theta$ (we drop the dependence of the coefficients on temperature in the following).

We introduce the back stresses $\boldsymbol{X}_1, \boldsymbol{X}_2$ and isotropic hardenings $R_1, R_2$ as thermodynamic forces via partial derivatives of the free energy with respect to the corresponding internal variables:

$$\boldsymbol{X}_j = \varrho\frac{\partial\psi_{kin}}{\partial\boldsymbol{\alpha}_j} \quad \text{and} \quad R_j = \varrho\frac{\partial\psi_{iso}}{\partial r_j} \quad, \quad j = 1, 2.$$
(3.2.36)

Thus, the definitions of the back stresses $\boldsymbol{X}_1, \boldsymbol{X}_2$ correspond to the one stated in (3.2.9) in the case of a 2M1C model. The two isotropic hardenings $R_1, R_2$ are given by

$$R_1 = \varrho\frac{\partial\psi_{in}}{\partial r_1} = d_{11}r_1 + d_{12}r_2$$
(3.2.37a)

$$R_2 = \varrho\frac{\partial\psi_{in}}{\partial r_2} = d_{12}r_1 + d_{22}r_2,$$
(3.2.37b)

using (3.2.36) and (3.2.34).

Now, the remaining inequality (3.1.6) can be specified as shown in Section 3.2.1.

Using (3.2.33), (3.2.34) and the definitions of the thermodynamic forces in (3.2.9), (3.2.37), we obtain:

$$\boldsymbol{\sigma}_1 : \dot{\boldsymbol{\varepsilon}}_1 + \boldsymbol{\sigma}_2 : \dot{\boldsymbol{\varepsilon}}_2 - \boldsymbol{X}_1 : \dot{\boldsymbol{\alpha}}_1 - \boldsymbol{X}_2 : \dot{\boldsymbol{\alpha}}_2 - R_1\dot{r}_1 - R_2\dot{r}_2 \geq 0.$$
(3.2.38)

This can be reformulated in

$$(\boldsymbol{\sigma}_1 - \boldsymbol{X}_1) : \dot{\boldsymbol{\varepsilon}}_1 + (\boldsymbol{\sigma}_2 - \boldsymbol{X}_2) : \dot{\boldsymbol{\varepsilon}}_2 + \boldsymbol{X}_1 : (\dot{\boldsymbol{\varepsilon}}_1 - \dot{\boldsymbol{\alpha}}_1) + \boldsymbol{X}_2 : (\dot{\boldsymbol{\varepsilon}}_2 - \dot{\boldsymbol{\alpha}}_2) +$$
$$- R_1\dot{r}_1 - R_2\dot{r}_2 \geq 0.$$
(3.2.39)

This form of the inequality will be used to verify the thermodynamic consistency of the model. Details will be given in the following subsection.

In the case of two mechanisms with two criteria, we have two *separate* yield functions $f_1, f_2$. We set:

$$f_j(\boldsymbol{\sigma}_j, \boldsymbol{X}_j, R_j, R_{0j}) := J_j - (R_j + R_{0j}) \,, \quad j = 1, 2 \,, \qquad (3.2.40)$$

where $J_j$ is defined as in (3.2.13). $R_{0j}$ corresponds to the initial yield stress of the $j^{th}$ mechanism ($j = 1, 2$). In the case of plastic mechanisms, these functions distinguish the elastic from the plastic domain. We suppose the constraints

$$f_j(\boldsymbol{\sigma}_j, \boldsymbol{X}_j, R_j, R_{0j}) \leq 0 \,, \qquad (3.2.41)$$

for $j = 1, 2$. Furthermore, we define

$$n_j := -\frac{\partial f_j}{\partial \boldsymbol{X}_j} = \sqrt{\frac{3}{2}} \frac{\boldsymbol{\sigma}_j^* - \boldsymbol{X}_j^*}{\|\boldsymbol{\sigma}_j^* - \boldsymbol{X}_j^*\|} = \frac{3}{2} \frac{\boldsymbol{\sigma}_j^* - \boldsymbol{X}_j^*}{J_j} \,, \qquad (3.2.42)$$

using the definition of $J_j$ in (3.2.13) and the yield function (3.2.40). Finally, we assume the following evolution laws for the inelastic strains introducing the two scalar plastic multipliers $\gamma_1, \gamma_2$:

$$\dot{\boldsymbol{\varepsilon}}_j = \gamma_j \, n_j \,, \quad j = 1, 2 \,. \qquad (3.2.43)$$

The plastic multipliers have to fulfil:

$$\gamma_j = 0 \quad \text{if} \quad f_j(\boldsymbol{\sigma}_j, \boldsymbol{X}_j, R_j, R_{0j}) < 0 \,, \qquad (3.2.44a)$$

$$\gamma_j \geq 0 \quad \text{if} \quad f_j(\boldsymbol{\sigma}_j, \boldsymbol{X}_j, R_j, R_{0j}) = 0 \,, \qquad (3.2.44b)$$

where $j = 1, 2$. Using the definitions of the accumulated inelastic strains (3.1.3) together with the definitions in (3.2.13), (3.2.42) and the evolution equations (3.2.43) for the inelastic strains, we obtain for the plastic multipliers

$$\gamma_j = \dot{s}_j \quad , \quad j = 1, 2 \,. \qquad (3.2.45)$$

For the scalar internal variables $r_1, r_2$ we set:

$$\dot{r}_j = \gamma_j - \sum_{i=1}^{2} a_{ji} R_i \sqrt{\gamma_j \gamma_i} \quad , \qquad j = 1, 2 \,, \qquad (3.2.46)$$

where we assume the matrix $\boldsymbol{a}$ consisting of the coefficients $a_{ji}$ in (3.2.46) to be positive semi-definite.

Furthermore, we define evolution laws for the tensorial internal variables $\boldsymbol{\alpha}_1, \boldsymbol{\alpha}_2$. We follow the extended approach presented in Wolff et al. (2011c) (cf. also Wolff et al. (2015)) which takes into account both multipliers for each variable[3]:

$$\dot{\boldsymbol{\alpha}}_j = \dot{\boldsymbol{\varepsilon}}_j - \frac{3}{2} \sum_{i=1}^{2} b_{ji} \boldsymbol{X}_i \sqrt{\gamma_j \gamma_i} \quad , \quad j = 1, 2 \,. \qquad (3.2.47)$$

Again, we assume that the matrix $\boldsymbol{b}$ composed of the coefficients $b_{ji}$ is positive semi-definite. Using (3.2.9) with (3.2.47), one obtains the generalised Armstrong-Frederick equations for the evolution of the back stresses $\boldsymbol{X}_1, \boldsymbol{X}_2$.

---

[3]The approach in (3.2.47) is an extension of the wide-spread approach where $b_{12} = b_{21} = 0$, cf. Wolff and Böhm (2010) and Chaboche (2008).

**Remark 3.2.5.** *Considering plasticity, the plastic multipliers $\gamma_j$ in (3.2.43) are determined via consistency conditions (see (3.2.44)). In the case of creep, we will take the specific material law of the corresponding creep mechanism into account which allows us to state the multiplier directly. In Section 3.3.2, we will present a 2M2C model for the modelling of creep together with the corresponding multipliers $\gamma_1, \gamma_2$.*

We refer to Wolff et al. (2010) and Wolff et al. (2011c) for further details about 2M2C models with plastic mechanisms.

**Thermodynamic consistency**

In order to prove the thermodynamic consistency of the presented 2M1C model, we have to verify the remaining inequality (3.2.39). Therefore, we insert (3.2.43), (3.2.47) and (3.2.46) in (3.2.39), which yields the inequality:

$$\gamma_1 (\boldsymbol{\sigma}_1 - \boldsymbol{X}_1) : n_1 + \gamma_2 (\boldsymbol{\sigma}_2 - \boldsymbol{X}_2) : n_2 +$$

$$+ \frac{3}{2} \sum_{i,j=1}^{2} b_{ij} \, \boldsymbol{X}_i : \boldsymbol{X}_j \, \sqrt{\gamma_i \gamma_j} - \sum_{i=1}^{2} R_i \gamma_i + \sum_{i,j=1}^{2} a_{ij} R_i R_j \sqrt{\gamma_i \gamma_j} \; \geq 0 \; .$$
$$(3.2.48)$$

Using definition (3.2.42) for $n_j$, we obtain

$$\gamma_j (\boldsymbol{\sigma}_j - \boldsymbol{X}_j) : n_j = \gamma_j J_j \quad , \quad \text{for} \quad j = 1,2 \; . \qquad (3.2.49)$$

Taking the definition of the yield functions $f_j$ in (3.2.40) together with the yield criterion $f_j = 0$ for $j = 1, 2$, in (3.2.44b) - otherwise the corresponding plastic multiplier is equal to zero - yields

$$J_j = R_j + R_{0j} \quad , \quad j = 1, 2 \; . \qquad (3.2.50)$$

Altogether, (3.2.48) can be rewritten into

$$\sum_{i=1}^{2} \gamma_i R_{0i} + \sum_{i,j=1}^{2} a_{ij} R_i R_j \sqrt{\gamma_i \gamma_j} + \frac{3}{2} \sum_{i,j=1}^{2} b_{ij} \, \boldsymbol{X}_i : \boldsymbol{X}_j \, \sqrt{\gamma_i \gamma_j} \; \geq 0 \; . \qquad (3.2.51)$$

Due to the non-negativity of $\gamma_1, \gamma_2$, the inequality (3.2.51) is fulfilled under the assumption that the matrices $\boldsymbol{a}, \boldsymbol{b}$ consisting of the coefficients $a_{ij}$ and $b_{ij}$, respectively, are positive semi-definite. This ensures the thermodynamic consistency of the presented model.

We refer to Wolff and Taleb (2008) for further details as well as to Wolff et al. (2010), Wolff et al. (2011c) and Wolff et al. (2015) for an extended version of the model requiring additional conditions in order to ensure thermodynamic consistency.

In the following Sections we will present some examples of 2M models and prove their thermodynamic consistency. In Section 3.3, we will present a model with creep mechanisms. Section 3.4 deals with a model with creep and TRIP mechanisms. See 3.3.3 and 3.4.2 for details about the thermodynamic consistency of the presented models.

## 3.3 Modelling of creep by means of two-mechanism models

In the following, we will focus on the modelling of creep using two-mechanism models. We consider our model presented in Chapter 2, see Section 2.4. We will exploit the fact that one is able to model creep similarly to the modelling of plasticity which was handled in Section 3.2. The presented models will be adapted to our case in the following. As in the case of creep there is no yield stress, we will set the yield stress formally equal to zero. Note that, in the case of creep, the multipliers $\gamma_i$ are not determined via consistency conditions but can be stated directly using the underlying material law. In order to provide the required model equations, we will formally state a yield function.

We will consider two different types of two-mechanism models: First, we will consider a model consisting of two mechanisms with one common criterion, second a model with two criteria (one for each mechanism), abbreviated as 2M1C and 2M2C model, respectively, cf. Sections 3.2.1 and 3.2.2.

In the following, the inelastic strain corresponds to the creep strain. It is decomposed into

$$\varepsilon_{in} = \varepsilon_c = A_1 \varepsilon_{c1} + A_2 \varepsilon_{c2} , \qquad (3.3.1)$$

where $A_1, A_2$ are real numbers and $A_1, A_2 > 0$ (cf. Remark 3.1.1). As introduced in Section 3.2, we define the accumulated creep strains for each creep mechanism $\varepsilon_{cj}$ by

$$s_{cj}(t) := \int_0^t \sqrt{\frac{2}{3} \dot{\varepsilon}_{cj}(\tau) : \dot{\varepsilon}_{cj}(\tau)} \, d\tau \quad , \quad j = 1, 2 . \qquad (3.3.2)$$

Next, we specify the corresponding model equations in the case of creep mechanisms. Note that, in contrast to plastic material behaviour, in the case of creep there is no yield stress. Therefore, only kinematic hardening is considered in the following.

### 3.3.1 Modelling of creep using a 2M1C model

We specialise the ansatz for the inelastic part of the free energy $\psi_{in}$ in (2.1.7). We assume that the internal variables are given as $\xi = (\alpha_1, \alpha_2)$, thus we have

$$\psi_{in} = \psi_{in}(\alpha_1, \alpha_2, \theta) . \qquad (3.3.3)$$

Here, $\alpha_1, \alpha_2$ are tensorial symmetric internal variables of strain type that are related to kinematic hardening. The internal variables $\alpha_j$ are each associated with the creep mechanisms $\varepsilon_{cj}$, $j = 1, 2$.

We use the definition of the inelastic part of the free energy in the case of a 2M1C model given in (3.2.4)-(3.2.6). In our case, the definition simplifies to

$$\psi_{in}(\alpha_1, \alpha_2, \theta) := \frac{1}{3\varrho}(c_{11}(\theta)\alpha_1 : \alpha_1 + 2c_{12}(\theta)\alpha_1 : \alpha_2 + c_{22}(\theta)\alpha_2 : \alpha_2) . \qquad (3.3.4)$$

Assuming

$$c_{11} > 0 \quad \text{and} \quad c_{12}^2(\theta) \leq c_{11}(\theta)c_{22}(\theta) \,, \qquad (3.3.5)$$

for all admissible temperatures $\theta$, the inelastic free energy $\psi_{in}$ is a convex function. In the following, we drop the dependence of the coefficients on temperature. The back stresses $\boldsymbol{X}_{c1}, \boldsymbol{X}_{c2}$ associated with the mechanisms $\boldsymbol{\varepsilon}_{c1}, \boldsymbol{\varepsilon}_{c2}$, respectively, are defined via the partial derivatives of the free energy with respect to the corresponding internal variables:

$$\boldsymbol{X}_{c1} := \varrho \frac{\partial \psi_{in}}{\partial \boldsymbol{\alpha}_1} = \frac{2}{3}c_{11}\boldsymbol{\alpha}_1 + \frac{2}{3}c_{12}\boldsymbol{\alpha}_2 \,, \qquad (3.3.6a)$$

$$\boldsymbol{X}_{c2} := \varrho \frac{\partial \psi_{in}}{\partial \boldsymbol{\alpha}_2} = \frac{2}{3}c_{12}\boldsymbol{\alpha}_1 + \frac{2}{3}c_{22}\boldsymbol{\alpha}_2 \,. \qquad (3.3.6b)$$

As no isotropic hardening arises, the remaining inequality in (3.2.11) reduces to

$$\boldsymbol{\sigma}_1 : \dot{\boldsymbol{\varepsilon}}_{c1} + \boldsymbol{\sigma}_2 : \dot{\boldsymbol{\varepsilon}}_{c2} - \boldsymbol{X}_{c1} : \dot{\boldsymbol{\alpha}}_1 - \boldsymbol{X}_{c2} : \dot{\boldsymbol{\alpha}}_2 \geq 0 \,, \qquad (3.3.7)$$

which can be written as

$$(\boldsymbol{\sigma}_1 - \boldsymbol{X}_{c1}) : \dot{\boldsymbol{\varepsilon}}_{c1} + (\boldsymbol{\sigma}_2 - \boldsymbol{X}_{c2}) : \dot{\boldsymbol{\varepsilon}}_{c2} + \boldsymbol{X}_{c1} : (\dot{\boldsymbol{\varepsilon}}_{c1} - \dot{\boldsymbol{\alpha}}_1) + \boldsymbol{X}_{c2} : (\dot{\boldsymbol{\varepsilon}}_{c2} - \dot{\boldsymbol{\alpha}}_2) \geq 0 \,. \qquad (3.3.8)$$

As introduced in Section 3.2.1, we use the definitions

$$J_j := \sqrt{\frac{3}{2}} \|\boldsymbol{\sigma}_j^* - \boldsymbol{X}_{cj}^*\| = \sqrt{\frac{3}{2}(\boldsymbol{\sigma}_j^* - \boldsymbol{X}_{cj}^*) : (\boldsymbol{\sigma}_j^* - \boldsymbol{X}_{cj}^*)} \,, \quad j = 1, 2 \,, \qquad (3.3.9)$$

and

$$J := \left( J_1^N + J_2^N \right)^{\frac{1}{N}} \,, \quad N > 1 \,. \qquad (3.3.10)$$

We follow the approach in Section 3.2.1. In the case of creep, there is no elastic domain and thus no yield stress. Nevertheless, we can formally define a yield function $f_c$ using (3.2.15) with $R = 0$ and setting $R_0$ formally equal to zero:

$$f_c(\boldsymbol{\sigma}_1, \boldsymbol{\sigma}_2, \boldsymbol{X}_{c1}, \boldsymbol{X}_{c1}) := J \,. \qquad (3.3.11)$$

Furthermore, we use $n_j$ as defined in (3.2.17), i.e.

$$n_j := -\frac{\partial f_c}{\partial \boldsymbol{X}_{cj}} = \frac{3}{2} \frac{\boldsymbol{\sigma}_j^* - \boldsymbol{X}_{cj}^*}{J_j} \left( \frac{J_j}{J} \right)^{N-1} \,, \quad j = 1, 2 \,, \qquad (3.3.12)$$

taking into account (3.3.9),(3.3.10) and (3.3.11).

Finally, we specify the evolution laws for the creep strains $\boldsymbol{\varepsilon}_{c1}, \boldsymbol{\varepsilon}_{c2}$. We are able to state evolution equations of the form (3.2.18) for the creep strains, too. Therefore, we introduce the common multiplier $\gamma_c$ and obtain material laws of the form

$$\dot{\boldsymbol{\varepsilon}}_{cj} = \gamma_c n_j \,, \quad j = 1, 2 \,. \qquad (3.3.13)$$

In the case of creep, the multiplier is not determined via a consistency condition (as in the case of classical plasticity, cf. (3.2.19) and Table 3.1). Contrary to elasto-plastic material behaviour, there is no elastic domain and the creep mechanisms are thus always active.

We will directly state the multiplier using the material law for creep. In the expression for $\gamma_c$, we will take both mechanisms into account.

First, we define

$$\Gamma_c(t) := \int_0^t \left( \dot{s}_{c1}^{\frac{N}{N-1}} + \dot{s}_{c2}^{\frac{N}{N-1}} \right)^{\frac{N-1}{N}} d\tau \,, \tag{3.3.14}$$

where $s_{cj}$ denote the accumulated strains, as defined in (3.3.2).

In order to formulate the evolution equation for $\boldsymbol{\varepsilon}_{cj}$, we use the material law for creep (2.2.30) with $J$ defined in (3.3.10) instead of the von Mises stress, taking the local stresses and back stresses into account, the normal direction $n_j$ as defined in (3.3.12) and the integral expression (3.3.14) which includes the accumulated strains of *both* mechanisms. We obtain

$$\dot{\boldsymbol{\varepsilon}}_{cj} = \frac{3}{2} A_c \left( \frac{J}{D_c} \right)^m \frac{\boldsymbol{\sigma}_j^* - \boldsymbol{X}_{cj}^*}{J_j} \left( \frac{J_j}{J} \right)^{N-1} \Gamma_c^k \,, \tag{3.3.15}$$

for $j = 1, 2$. By setting

$$\gamma_c = A_c \left( \frac{J}{D_c} \right)^m \Gamma_c^k \,, \tag{3.3.16}$$

for the common multiplier and using $n_j$ defined in (3.3.12), we have

$$\dot{\boldsymbol{\varepsilon}}_{cj} = \gamma_c \, n_j \quad , \quad j = 1, 2 \,, \tag{3.3.17}$$

where the multiplier $\gamma_c$ is always non-negative. Furthermore, using (3.3.14), (3.3.15) and (3.3.17) yields

$$\gamma_c = \left( \dot{s}_1^{\frac{N}{N-1}} + \dot{s}_2^{\frac{N}{N-1}} \right)^{\frac{N-1}{N}} \,. \tag{3.3.18}$$

Thus, equation (3.3.16) is fulfilled and

$$\Gamma_c(t) = \int_0^t \gamma_c(\tau) \, d\tau \,. \tag{3.3.19}$$

In the next step, we provide evolution laws for the internal variables $\boldsymbol{\alpha}_1, \boldsymbol{\alpha}_2$. We set

$$\dot{\boldsymbol{\alpha}}_j = \dot{\boldsymbol{\varepsilon}}_{cj} - \frac{3}{2} \sum_{i=1}^2 b_{ji} \boldsymbol{X}_{c_i} \gamma_c \quad , \quad j = 1, 2 \,, \tag{3.3.20}$$

taking the approaches Wolff and Böhm (2010), Wolff et al. (2011c), Wolff et al. (2013) into account. The above approach is an extension of the wide-spread approach where $b_{12} = b_{21} = 0$, cf. Wolff and Böhm (2010) and Chaboche (2008). For constant $c_{ij}$, (3.3.6) and (3.3.20) yield the generalised Armstrong-Frederick equations for the evolution of the back stresses:

$$\dot{\boldsymbol{X}}_{c1} = \frac{2}{3}c_{11}\dot{\boldsymbol{\varepsilon}}_{c1} - c_{11}(b_{11}\boldsymbol{X}_{c1} + b_{12}\boldsymbol{X}_{c2})\gamma_c + \frac{2}{3}c_{12}\dot{\boldsymbol{\varepsilon}}_{c2} - c_{12}(b_{21}\boldsymbol{X}_{c1} + b_{22}\boldsymbol{X}_{c2})\gamma_c \,,$$
(3.3.21a)

$$\dot{\boldsymbol{X}}_{c2} = \frac{2}{3}c_{12}\dot{\boldsymbol{\varepsilon}}_{c1} - c_{12}(b_{11}\boldsymbol{X}_{c1} + b_{12}\boldsymbol{X}_{c2})\gamma_c + \frac{2}{3}c_{22}\dot{\boldsymbol{\varepsilon}}_{c2} - c_{22}(b_{21}\boldsymbol{X}_{c1} + b_{22}\boldsymbol{X}_{c2})\gamma_c \,.$$
(3.3.21b)

Under the assumption that the matrix $\boldsymbol{b}$ consisting of the parameters $b_{ji}, i, j = 1, 2$ in (3.3.20) is positive semi-definit, the presented model is thermodynamically consistent for creep mechanisms. The thermodynamic consistency of the model can be shown by verifying (3.3.8). More details will be given in the following Section 3.3.3.

### 3.3.2 Modelling of creep using a 2M2C model

We specialise the ansatz for the inelastic part of the free energy $\psi_{in}$ (cf. Section 3.2.2). We assume the internal variables to be given as $\xi = (\boldsymbol{\alpha}_1, \boldsymbol{\alpha}_2)$, thus

$$\psi_{in} = \psi_{in}(\boldsymbol{\alpha}_1, \boldsymbol{\alpha}_2, \theta) \,.$$
(3.3.22)

Here, $\boldsymbol{\alpha}_1, \boldsymbol{\alpha}_2$ are tensorial symmetric internal variables of strain type that are related to kinematic hardening. The internal variables $\boldsymbol{\alpha}_j$ are each associated with mechanism $\boldsymbol{\varepsilon}_{cj}, j = 1, 2$.

Next, we define the inelastic part of the free energy for a 2M2C model according to (3.2.32)-(3.2.34). In the case of creep, this definition simplifies to

$$\psi_{in}(\boldsymbol{\alpha}_1, \boldsymbol{\alpha}_2, \theta) := \frac{1}{3\varrho}(c_{11}(\theta)\boldsymbol{\alpha}_1 : \boldsymbol{\alpha}_1 + 2c_{12}(\theta)\boldsymbol{\alpha}_1 : \boldsymbol{\alpha}_2 + c_{22}(\theta)\boldsymbol{\alpha}_2 : \boldsymbol{\alpha}_2) \,,$$
(3.3.23)

which is equal to equation (3.3.4) of $\psi_{in}$ in the case of a 2M1C model.

Assuming

$$c_{11} > 0 \quad \text{and} \quad c_{12}^2(\theta) \le c_{11}(\theta)c_{22}(\theta) \,,$$
(3.3.24)

for all admissible temperatures $\theta$, the inelastic free energy $\psi_{in}$ is a convex function.

The back stresse s $\boldsymbol{X}_{c1}, \boldsymbol{X}_{c2}$ associated with the mechanisms $\boldsymbol{\varepsilon}_{c1}, \boldsymbol{\varepsilon}_{c2}$, respectively, are defined via the partial derivatives of the free energy with respect to the corresponding internal variable:

$$\boldsymbol{X}_{c1} := \varrho\frac{\partial\psi_{in}}{\partial\boldsymbol{\alpha}_1} = \frac{2}{3}c_{11}\boldsymbol{\alpha}_1 + \frac{2}{3}c_{12}\boldsymbol{\alpha}_2 \,,$$
(3.3.25a)

$$\boldsymbol{X}_{c2} := \varrho\frac{\partial\psi_{in}}{\partial\boldsymbol{\alpha}_2} = \frac{2}{3}c_{12}\boldsymbol{\alpha}_1 + \frac{2}{3}c_{22}\boldsymbol{\alpha}_2 \,.$$
(3.3.25b)

The remaining inequality (3.2.38) reduces to

$$\boldsymbol{\sigma}_1 : \dot{\boldsymbol{\varepsilon}}_{c1} + \boldsymbol{\sigma}_2 : \dot{\boldsymbol{\varepsilon}}_{c2} - \boldsymbol{X}_{c1} : \dot{\boldsymbol{\alpha}}_1 - \boldsymbol{X}_{c2} : \dot{\boldsymbol{\alpha}}_2 \ge 0 \,.$$
(3.3.26)

We rewrite this equation in the form

$$(\boldsymbol{\sigma}_1 - \boldsymbol{X}_{c1}) : \dot{\boldsymbol{\varepsilon}}_{c1} + (\boldsymbol{\sigma}_2 - \boldsymbol{X}_{c2}) : \dot{\boldsymbol{\varepsilon}}_{c2} + \boldsymbol{X}_{c1} : (\dot{\boldsymbol{\varepsilon}}_{c1} - \dot{\boldsymbol{\alpha}}_1) + \boldsymbol{X}_{c2} : (\dot{\boldsymbol{\varepsilon}}_{c2} - \dot{\boldsymbol{\alpha}}_2) \geq 0 . \tag{3.3.27}$$

So far, there is no difference to the modelling of creep using a 2M1C model.

In the next step, the definition of yield functions is required. Note that there is no yield stress in the case of creep. However, one can introduce yield functions $f_{c1}, f_{c2}$ by formally setting $R_1 = R_2 = 0$ and $R_{01} = R_{02} = 0$. We use (3.2.40) and set for the two (formal) yield functions

$$f_{cj}(\boldsymbol{\sigma}_j, \boldsymbol{X}_{cj}, R_{0j}) := J_j , \tag{3.3.28}$$

where $J_j$ is defined by (3.3.9) and $j = 1, 2$. Furthermore, we have

$$n_j := -\frac{\partial f_{cj}}{\partial \boldsymbol{X}_{cj}} = \sqrt{\frac{3}{2}} \frac{\boldsymbol{\sigma}_j^* - \boldsymbol{X}_{cj}^*}{\|\boldsymbol{\sigma}_j^* - \boldsymbol{X}_{cj}^*\|} = \frac{3}{2} \frac{\boldsymbol{\sigma}_j^* - \boldsymbol{X}_{cj}^*}{J_j} , \tag{3.3.29}$$

taking into account (3.3.28) and (3.3.9). Finally, we specify the evolution laws for the creep strains $\boldsymbol{\varepsilon}_{c1}, \boldsymbol{\varepsilon}_{c2}$. If the $j^{th}$ mechanism models creep, we use the following evolution equation for the corresponding creep strain, following the material law for creep introduced in Section 2.2.2:

$$\dot{\boldsymbol{\varepsilon}}_{cj} = \frac{3}{2} A_{cj} \left( \sqrt{\frac{3}{2}} \frac{\|\boldsymbol{\sigma}_j^* - \boldsymbol{X}_{cj}^*\|}{D_{cj}} \right)^{m_j - 1} \frac{\boldsymbol{\sigma}_j^* - \boldsymbol{X}_{cj}^*}{D_{cj}} s_{cj}^{k_j} , \quad j = 1, 2, \tag{3.3.30}$$

with the parameters $A_{cj} > 0$, $m_j > 0$, $k_j$ that depend on the temperature $\theta$ and the drag stress $D_{cj} > 0$. The parameter $k_j$ distinguishes the different creep stages (cf. (2.2.32)). The drag stress $D_{cj}$ may be either constant or have an own evolution equation. See Section 2.2.2 for further details.

As mentioned above, in the case of creep, the multipliers are not determined via consistency conditions. The creep mechanisms are always active. In order to formulate general evolution laws of the form (3.2.43) for the creep strains, i.e.

$$\dot{\boldsymbol{\varepsilon}}_{cj} = \gamma_{cj} n_j , \quad j = 1, 2 , \tag{3.3.31}$$

we define the inelastic scalar multipliers $\gamma_{c1}, \gamma_{c2}$ by setting

$$\gamma_{cj} := A_{cj} \left( \sqrt{\frac{3}{2}} \frac{\|\boldsymbol{\sigma}_j^* - \boldsymbol{X}_{cj}^*\|}{D_{cj}} \right)^{m_j} s_{cj}^{k_j} , \quad j = 1, 2 . \tag{3.3.32}$$

Using (3.3.29), (3.3.30) and (3.3.32), we obtain

$$\dot{\boldsymbol{\varepsilon}}_{cj} = \frac{3}{2} A_{cj} \left( \sqrt{\frac{3}{2}} \frac{\|\boldsymbol{\sigma}_j^* - \boldsymbol{X}_{cj}^*\|}{D_{cj}} \right)^{m_j - 1} \frac{\boldsymbol{\sigma}_j^* - \boldsymbol{X}_{cj}^*}{D_{cj}} s_{cj}^{k_j} = \gamma_{cj} n_j , \quad j = 1, 2 . \tag{3.3.33}$$

In the next step, we define evolution laws for the internal variables $\boldsymbol{\alpha}_1, \boldsymbol{\alpha}_2$. We follow the approach in (3.2.47):

$$\dot{\boldsymbol{\alpha}}_j = \dot{\boldsymbol{\varepsilon}}_{cj} - \frac{3}{2} \sum_{i=1}^{2} b_{ji} \boldsymbol{X}_{ci} \sqrt{\gamma_{cj}\gamma_{ci}} \quad, \quad j = 1, 2 . \tag{3.3.34}$$

The approach in (3.3.34) is an extension of the wide-spread approach where $b_{12} = b_{21} = 0$, cf. Wolff and Böhm (2010) and Chaboche (2008). For constant $c_{ij}, i, j = 1, 2$, (3.3.25) and (3.3.34) yield the generalised Armstrong-Frederick equations for the evolution of the back stresses:

$$\dot{\boldsymbol{X}}_{c1} = c_{11} \left( \frac{2}{3} \dot{\boldsymbol{\varepsilon}}_{c1} - \left( b_{11} \boldsymbol{X}_{c1} \gamma_{c1} + b_{12} \boldsymbol{X}_{c2} \sqrt{\gamma_{c1}\gamma_{c2}} \right) \right) +$$
$$+ \; c_{12} \left( \frac{2}{3} \dot{\boldsymbol{\varepsilon}}_{c2} - \left( b_{21} \boldsymbol{X}_{c1} \sqrt{\gamma_{c2}\gamma_{c1}} + b_{22} \boldsymbol{X}_{c2} \gamma_{c2} \right) \right) , \tag{3.3.35a}$$

$$\dot{\boldsymbol{X}}_{c2} = c_{12} \left( \frac{2}{3} \dot{\boldsymbol{\varepsilon}}_{c1} - \left( b_{11} \boldsymbol{X}_{c1} \gamma_{c1} + b_{12} \boldsymbol{X}_{c2} \sqrt{\gamma_{c1}\gamma_{c2}} \right) \right) +$$
$$+ \; c_{22} \left( \frac{2}{3} \dot{\boldsymbol{\varepsilon}}_{c2} - \left( b_{21} \boldsymbol{X}_{c1} \sqrt{\gamma_{c2}\gamma_{c1}} + b_{22} \boldsymbol{X}_{c2} \gamma_{c2} \right) \right) . \tag{3.3.35b}$$

In order to ensure thermodynamic consistency, we require that the matrix $\boldsymbol{b}$ consisting of the parameters $b_{ji}$, $i, j = 1, 2$ in equation (3.3.34) is positive semi-definite. Under this assumption, the 2M model with creep mechanisms presented above is thermodynamically consistent. The thermodynamic consistency of the model can be shown using (3.3.27). Details will be given in Section 3.3.3.

### 3.3.3 Thermodynamic consistency

Now, we investigate the thermodynamic consistency of the presented models. In order to ensure that a model is thermodynamically consistent, we have to verify the corresponding remaining inequality (2.1.19).

**Two-mechanism model with one criterion**

In the case of a 2M1C model, we have a closer look at the reduced dissipation inequality given in (3.3.8):

$$(\boldsymbol{\sigma}_1 - \boldsymbol{X}_{c1}) : \dot{\boldsymbol{\varepsilon}}_{c1} + (\boldsymbol{\sigma}_2 - \boldsymbol{X}_{c2}) : \dot{\boldsymbol{\varepsilon}}_{c2} + \boldsymbol{X}_{c1} : (\dot{\boldsymbol{\varepsilon}}_{c1} - \dot{\boldsymbol{\alpha}}_1) + \boldsymbol{X}_{c2} : (\dot{\boldsymbol{\varepsilon}}_{c2} - \dot{\boldsymbol{\alpha}}_2) \geq 0 . \tag{3.3.36}$$

First, we consider the first two parts of the sum stated above.

The corresponding material law for $\boldsymbol{\varepsilon}_{cj}$ is

$$\dot{\boldsymbol{\varepsilon}}_{cj} = \gamma_c \, n_j , \quad j = 1, 2, \tag{3.3.37}$$

where the common multiplier is $\gamma_c \geq 0$ (cf. (3.3.16)). We insert the material law in the remaining inequality and obtain for the first two parts of the sum in (3.3.36):

$$\gamma_c (\boldsymbol{\sigma}_1 - \boldsymbol{X}_{c1}) : n_1 + \gamma_c (\boldsymbol{\sigma}_2 - \boldsymbol{X}_{c2}) : n_2 . \tag{3.3.38}$$

Using the definition of $n_j$ (3.3.12), the definitions (3.2.13), (3.2.14) and the fact that for a tensor $\boldsymbol{\tau}$ and its deviator $\boldsymbol{\tau}^*$ holds that:

$$\boldsymbol{\tau} : \boldsymbol{\tau}^* = \boldsymbol{\tau}^* : \boldsymbol{\tau}^* , \tag{3.3.39}$$

we obtain:

$$\begin{aligned}(\boldsymbol{\sigma}_j - \boldsymbol{X}_{cj}) : \dot{\boldsymbol{\varepsilon}}_{cj} &= \frac{3}{2} \gamma_c \frac{(\boldsymbol{\sigma}_j - \boldsymbol{X}_{cj}) : (\boldsymbol{\sigma}_j^* - \boldsymbol{X}_{cj}^*)}{J_j} \left(\frac{J_j}{J}\right)^{N-1} \\ &= \gamma_c J_j^N \frac{1}{J^{N-1}} \end{aligned} \tag{3.3.40}$$

with $j = 1, 2$. Summing both parts up, we get

$$(\boldsymbol{\sigma}_1 - \boldsymbol{X}_{c1}) : \dot{\boldsymbol{\varepsilon}}_{c1} + (\boldsymbol{\sigma}_2 - \boldsymbol{X}_{c2}) : \dot{\boldsymbol{\varepsilon}}_{c2} = \gamma_c J \tag{3.3.41}$$

which is non-negative.

Next, we consider the last two parts of the sum in (3.3.36). We insert the evolution equation for $\boldsymbol{\alpha}_j$ as stated in (3.3.20) and obtain:

$$\begin{aligned}\boldsymbol{X}_{c1} : (\dot{\boldsymbol{\varepsilon}}_{c1} - \dot{\boldsymbol{\alpha}}_1) &+ \boldsymbol{X}_{c2} : (\dot{\boldsymbol{\varepsilon}}_{c2} - \dot{\boldsymbol{\alpha}}_2) = \\ &= \frac{3}{2}(b_{11}\boldsymbol{X}_{c1} : \boldsymbol{X}_{c1} + b_{12}\boldsymbol{X}_{c2} : \boldsymbol{X}_{c1})\gamma_c + \frac{3}{2}(b_{21}\boldsymbol{X}_{c1} : \boldsymbol{X}_{c2} + b_{22}\boldsymbol{X}_{c2} : \boldsymbol{X}_{c2})\gamma_c \\ &= \frac{3}{2} \sum_{i,j=1,2} b_{ji}(\boldsymbol{X}_{ci} : \boldsymbol{X}_{cj})\gamma_c . \end{aligned} \tag{3.3.42}$$

This part of the remaining inequality is non-negative, if the matrix $\boldsymbol{b}$ consisting of the $b_{ji}$ in (3.3.42) is positive semi-definite. As a result, under this assumption (3.3.36) is fulfilled and thermodynamic consistency is ensured.

**Two-mechanism model with two criteria**

Now, we consider a 2M2C model with creep mechanisms as presented in Section 3.3.2. The reduced dissipation inequality in the case of a 2M2C model corresponds to the inequality (3.3.36) stated above for a 2M1C model.

First, we consider the first part of (3.3.36) and check if

$$(\boldsymbol{\sigma}_1 - \boldsymbol{X}_{c1}) : \dot{\boldsymbol{\varepsilon}}_{c1} + (\boldsymbol{\sigma}_2 - \boldsymbol{X}_{c2}) : \dot{\boldsymbol{\varepsilon}}_{c2} \geq 0 \tag{3.3.43}$$

is fulfilled. We use the material laws for the creep strains $\boldsymbol{\varepsilon}_{cj}$ which are given by

$$\dot{\boldsymbol{\varepsilon}}_{cj} = \gamma_{cj} \, n_j , \tag{3.3.44}$$

where $\gamma_{cj} \geq 0$ for $j = 1, 2$ (cf. (3.3.32)).

We insert the material laws in the remaining inequality. As above in the case of a 2M1C model, we exploit (3.3.39) and insert the definition (3.3.29) of $n_j$. We obtain for the first two parts of the sum in (3.3.36):

$$
\begin{aligned}
(\boldsymbol{\sigma}_j - \boldsymbol{X}_{cj}) : \dot{\boldsymbol{\varepsilon}}_{cj} &= \gamma_{cj}(\boldsymbol{\sigma}_j - \boldsymbol{X}_{cj}) : n_j \\
&= \frac{3}{2}\gamma_{cj}\, \frac{(\boldsymbol{\sigma}_j - \boldsymbol{X}_{cj}) : (\boldsymbol{\sigma}_j^* - \boldsymbol{X}_{cj}^*)}{J_j} \\
&= \gamma_{cj}\, J_j\,,
\end{aligned}
\tag{3.3.45}
$$

with $j = 1, 2$. Using the definitions of $J_j$ and $J$ (3.2.13), (3.2.14), respectively, we obtain for the sum (3.3.43)

$$
(\boldsymbol{\sigma}_1 - \boldsymbol{X}_{c1}) : \dot{\boldsymbol{\varepsilon}}_{c1} + (\boldsymbol{\sigma}_2 - \boldsymbol{X}_{c2}) : \dot{\boldsymbol{\varepsilon}}_{c2} = \gamma_{c1}J_1 + \gamma_{c2}J_2\,.
\tag{3.3.46}
$$

Thus, this part of (3.3.36) is non-negative.

In the same way as presented above, we handle the second part. We insert the evolution equations of $\boldsymbol{\alpha}_1, \boldsymbol{\alpha}_2$ (3.3.34). Thermodynamic consistency is ensured if it holds that

$$
\begin{aligned}
\boldsymbol{X}_{c1} : (\dot{\boldsymbol{\varepsilon}}_{c1} - \dot{\boldsymbol{\alpha}}_1) + \boldsymbol{X}_{c2} : (\dot{\boldsymbol{\varepsilon}}_{c2} - \dot{\boldsymbol{\alpha}}_2) &= \\
&= \frac{3}{2} \sum_{i,j=1,2} b_{ji}(\boldsymbol{X}_{ci} : \boldsymbol{X}_{cj})\sqrt{\gamma_{ci}\gamma_{cj}} \geq 0\,.
\end{aligned}
\tag{3.3.47}
$$

This is fulfilled if the matrix $\boldsymbol{b}$ consisting of the $b_{ji}$ in (3.3.47) is positive semi-definite.

Under this assumption, the remaining inequality holds and the 2M2C model is thermodynamically consistent.

**Remark 3.3.1.** *Further investigations considering thermodynamical consistency of multi-mechanism models can be found in* Wolff and Taleb (2008), Wolff et al. (2008), Wolff et al. (2010), Wolff et al. (2011c) *and* Kröger (2013).

## 3.4 A two-mechanism model for the modelling of creep and TRIP

Besides classical plasticity and creep, also other material laws for the single mechanisms in the MM model are possible.

further material behaviour for the single mechanisms of an MM model is possible. In Wolff et al. (2011b), the authors present a model taking interaction between classical plasticity and transformation-induced plasticity into account.

Here, we will present a similar model approach: We consider creep and transformation induced plasticity which occur in a material at the same time. The model which we will present in the following is a two-mechanism model with two criteria (cf. Section 3.2.2). The inelastic strain tensor is decomposed into two components:

$$
\boldsymbol{\varepsilon}_{in} = \boldsymbol{\varepsilon}_c + \boldsymbol{\varepsilon}_{trip}\,.
\tag{3.4.1}
$$

$\varepsilon_c$ is caused by creep, $\varepsilon_{trip}$ is caused by TRIP.

The model which will be considered in the following, is one example of a 2M2C model which was presented in Section 3.2.2. The partial inelastic strains are independent from each other, but coupled via their back stresses (see Section 3.4.1).

For the partial stresses in (3.2.2), it follows that $\boldsymbol{\sigma}_1 = \boldsymbol{\sigma}_2 = \boldsymbol{\sigma}$.

### 3.4.1 Evolution equations

First, we define the internal variables in (2.1.9) as

$$\xi = (\boldsymbol{\alpha}_c, \boldsymbol{\alpha}_{trip}) , \tag{3.4.2}$$

where $\boldsymbol{\alpha}_c, \boldsymbol{\alpha}_{trip}$ stand for the tensorial symmetric internal variables of strain type corresponding to the creep and the TRIP mechanism, respectively. Here, we do not consider a dependency of the free energy $\psi$ on the phase fractions $p$. We have

$$\psi_{in} = \psi_{in}(\xi, \theta) = \psi_{in}(\boldsymbol{\alpha}_c, \boldsymbol{\alpha}_{trip}, \theta) , \tag{3.4.3}$$

for the inelastic part of the free energy $\psi$ in (2.1.7) and define

$$\psi_{in}(\boldsymbol{\alpha}_c, \boldsymbol{\alpha}_{trip}, \theta) := \frac{1}{3\varrho}(c_{11}(\theta)\boldsymbol{\alpha}_c : \boldsymbol{\alpha}_c + 2c_{12}(\theta)\boldsymbol{\alpha}_c : \boldsymbol{\alpha}_{trip} + c_{22}(\theta)\boldsymbol{\alpha}_{trip} : \boldsymbol{\alpha}_{trip}) . \tag{3.4.4}$$

Assuming

$$c_{11} > 0 \quad \text{and} \quad c_{12}^2(\theta) \le c_{11}(\theta)c_{22}(\theta) , \tag{3.4.5}$$

for all admissible temperatures $\theta$, the inelastic free energy $\psi_{in}$ is a convex function.

As mentioned above, the single back stresses $\boldsymbol{X}_c$ and $\boldsymbol{X}_{trip}$ associated with the mechanisms $\varepsilon_c, \varepsilon_{trip}$, respectively, are coupled. In detail, the back stresses are defined via the partial derivatives of the free energy as

$$\boldsymbol{X}_c := \varrho \frac{\partial \psi_{in}}{\partial \boldsymbol{\alpha}_c} \quad , \quad \boldsymbol{X}_{trip} := \varrho \frac{\partial \psi_{in}}{\partial \boldsymbol{\alpha}_{trip}} \quad . \tag{3.4.6}$$

The special feature of this model lies in the fact that there is a *coupling* between the back stresses of the single inelastic strains.

From (3.4.4) and (3.4.6), we obtain

$$\boldsymbol{X}_c = \frac{2}{3}c_{11}\boldsymbol{\alpha}_c + \frac{2}{3}c_{12}\boldsymbol{\alpha}_{trip} , \tag{3.4.7}$$

$$\boldsymbol{X}_{trip} = \frac{2}{3}c_{12}\boldsymbol{\alpha}_c + \frac{2}{3}c_{22}\boldsymbol{\alpha}_{trip} . \tag{3.4.8}$$

The inelastic material behaviour is described via the following evolution equations (cf. Sections 2.2.2, 2.2.3 in Chapter 2). We use the material law for creep and introduce the scalar multiplier $\gamma_c$ as stated in (3.3.30) and (3.3.32), respectively,

$$\dot{\boldsymbol{\varepsilon}}_c = \frac{3}{2}A_c \left( \sqrt{\frac{3}{2}} \frac{\|\boldsymbol{\sigma}^* - \boldsymbol{X}_c^*\|}{D_c} \right)^{m-1} \frac{\boldsymbol{\sigma}^* - \boldsymbol{X}_c^*}{D_c} s_c^k = \sqrt{\frac{3}{2}}\gamma_c \frac{\boldsymbol{\sigma}^* - \boldsymbol{X}_c^*}{\|\boldsymbol{\sigma}^* - \boldsymbol{X}_c^*\|} , \tag{3.4.9}$$

where $A_c > 0, m > 0$ and the accumulated creep strain $s_c$ is as defined in (2.2.31). For the TRIP strain, we use the material law as stated in Chapter 2:

$$\dot{\varepsilon}_{trip} = \frac{3}{2} \left( \sigma^* - X^*_{trip} \right) \sum_{i=1}^{M_p} \kappa_i \frac{d\phi_i}{dp_i} \max\{ \dot{p}_i, 0 \}, \qquad (3.4.10)$$

with the Greenwood-Johnson parameter $\kappa_i > 0$ and a saturation function $\phi_i$ of the $i$-th phase $p_i$ (see Section 2.2.3). We model the TRIP strain in (3.4.10) with a back stress $X_{trip}$ (see Remark 2.2.3). For further explanations, we refer to Wolff et al. (2009), Wolff et al. (2011b), Wolff et al. (2011c), e.g.

For the internal variables $\alpha_c, \alpha_{trip}$ we set

$$\dot{\alpha}_c := \dot{\varepsilon}_c - \frac{3}{2} b_c X_c \dot{s}_c, \qquad (3.4.11)$$

and

$$\dot{\alpha}_{trip} := \dot{\varepsilon}_{trip} - \frac{3}{2} b_{trip} X_{trip} \dot{s}_{trip}. \qquad (3.4.12)$$

Using (3.4.7),(3.4.8),(3.4.11) and (3.4.12), we obtain the following evolution equations for the coupled back stresses:

$$\dot{X}_c = \frac{2}{3} c_{11} \dot{\varepsilon}_c - c_{11} b_c X_c \dot{s}_c + \frac{2}{3} c_{12} \dot{\varepsilon}_{trip} - c_{12} b_{trip} X_{trip} \dot{s}_{trip} \qquad (3.4.13)$$

$$\dot{X}_{trip} = \frac{2}{3} c_{12} \dot{\varepsilon}_c - c_{12} b_c X_c \dot{s}_c + \frac{2}{3} c_{22} \dot{\varepsilon}_{trip} - c_{22} b_{trip} X_{trip} \dot{s}_{trip}. \qquad (3.4.14)$$

**Remark 3.4.1** (3M models, further material behaviour).

*(i) Besides the model presented here, other types of multi-mechanism models are possible for the presented problem. For instance, the creep strain can be considered as one mechanism or as two mechanisms. In the latter case, the resulting model is a three-mechanism model with either two or three criteria. Considering one common creep multiplier $\gamma_c$, the resulting model is of 3M2C type. If each creep strain is modelled with an own multiplier, we have a 3M3C model.*

*As above, the TRIP and the creep strains are independent from each other but are coupled via their back stresses.*

*(ii) Furthermore, an extended version of the model presented above is possible, too. Additionally to creep and TRIP material behaviour, we can also consider classical plasticity. This can be modelled via a three-mechanism model. More precisely, in this case we would have a 3M3C model where the inelastic strain tensor is given by*

$$\varepsilon_{in} = \varepsilon_c + \varepsilon_{trip} + \varepsilon_{cp}. \qquad (3.4.15)$$

*The creep and the TRIP strain are modelled with a yield stress equal to zero, i.e. $R_{0c} = R_{0trip} = 0$ whereas for the yield stress of the plastic mechanism $\varepsilon_{cp}$ holds $R_{0cp} \neq 0$ [4]. The three inelastic strains are coupled with each other via their back stresses.*

---

[4]Besides this, also classical plasticity with a yield stress $R_{0cp} = 0$ is possible, see Haupt (2002).

## 3.4.2 Thermodynamic consistency

In order to ensure the thermodynamic consistency of the model, we consider the reduced remaining inequality (3.2.38). Thus, we have to verify

$$(\boldsymbol{\sigma} - \boldsymbol{X}_c) : \dot{\boldsymbol{\varepsilon}}_c + (\boldsymbol{\sigma} - \boldsymbol{X}_{trip}) : \dot{\boldsymbol{\varepsilon}}_{trip} + \boldsymbol{X}_c : (\dot{\boldsymbol{\varepsilon}}_c - \dot{\boldsymbol{\alpha}}_c) + \boldsymbol{X}_{trip} : (\dot{\boldsymbol{\varepsilon}}_{trip} - \dot{\boldsymbol{\alpha}}_{trip}) \geq 0 \,. \tag{3.4.16}$$

Considering the first part of (3.4.16), we use the material laws for $\boldsymbol{\varepsilon}_c$ and $\boldsymbol{\varepsilon}_{trip}$, (3.4.9) and (3.4.10), respectively, as well as formula (3.3.39). We obtain:

$$\sqrt{\frac{3}{2}}\, \gamma_c \frac{(\boldsymbol{\sigma} - \boldsymbol{X}_c) : (\boldsymbol{\sigma}^* - \boldsymbol{X}_c^*)}{\|\boldsymbol{\sigma}^* - \boldsymbol{X}_c^*\|} + \frac{3}{2}\, (\boldsymbol{\sigma} - \boldsymbol{X}_{trip}) : (\boldsymbol{\sigma}^* - \boldsymbol{X}_{trip}^*) \sum_{i=1}^{M_p} \kappa_i \frac{\mathrm{d}\phi_i}{\mathrm{d}p_i}\, \max\{\dot{p}_i, 0\}$$

$$= \sqrt{\frac{3}{2}}\, \gamma_c + \frac{3}{2}\, \|\boldsymbol{\sigma}^* - \boldsymbol{X}_{trip}^*\| \sum_{i=1}^{M_p} \kappa_i \frac{\mathrm{d}\phi_i}{\mathrm{d}p_i}\, \max\{\dot{p}_i, 0\} \geq 0 \,,$$

$$\tag{3.4.17}$$

due to the assumptions $\kappa_i > 0$ and $\frac{\mathrm{d}\phi_i}{\mathrm{d}p_i} \geq 0$, cf. Section 2.2.3.

Next, we insert the evolution equations of $\boldsymbol{\alpha}_c$ and $\boldsymbol{\alpha}_{trip}$, (3.4.11) and (3.4.12), in the second part of (3.4.16). In order to ensure the validity of (3.4.16), we require that

$$b_c \geq 0 \quad \text{and} \quad b_{trip} \geq 0 \,. \tag{3.4.18}$$

Altogether, the presented model is thermodynamically consistent under the stated assumptions.

# 4 Model verification and parameter identification

In this chapter, we aim to determine the concrete material behaviour of creep and TRIP by the verification of the assumed material laws and the determination of material parameters. We consider the corresponding material laws introduced in Chapter 2 and provide the related one-dimensional models.

First, in Section 4.1, we will present the 3D model for a one-dimensional stress state and derive the 1D versions of our model equations. In Section 4.2, we deal with the processing of experiments by deriving further data. After this, we present a general algorithm in Section 4.3 for the verification of material laws and apply this algorithm to the concrete material behaviour of creep and TRIP in Sections 4.4 and 4.5, respectively. In the considered material laws, the knowledge of certain material parameters is required. These parameters are determined via a parameter identification by means of appropriate experimental data. We deal with this topic in Section 4.6. We provide formulas for deriving certain data from uniaxial experiments and present the strategy for the identification of parameters.

We refer to Wolff et al. (2012c) for further details about the 1D algorithm and experimental results.

## 4.1 Preliminaries

We consider data from uniaxial experiments with small cylindrical specimen. Figure 4.1 illustrates a schematic representation of the specimen.

In uniaxial experiments with small specimen, spatial homogeneity over the gauge length is assumed. Hence, all measured and calculated quantities are only functions of time $t$. Section 4.2 deals with the given experimental data and describes how to derive further quantities which will be used in the parameter identification.

For further information, see Bökenheide and Wolff (2012), Wolff et al. (2012c), Wolff et al. (2012a), Wolff et al. (2011a).

### 4.1.1 Three-dimensional setting for a one-dimensional stress state

We consider the one-dimensional setting of a tension test with an applied stress $S$ in $x$-direction. First, we define the following scalar values referring to the entries

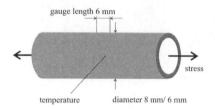

gauge length 6 mm

stress

temperature          diameter 8 mm/ 6 mm

Figure 4.1: Scheme of a hollow specimen used in the testing device Gleeble®. The measured data are length, diameter and stress as discrete functions of time.

of the corresponding tensors $\boldsymbol{\sigma}$ and $\boldsymbol{\varepsilon}$, respectively:

$$S := \sigma_{11} \quad \text{and} \quad \epsilon_{in} := \varepsilon_{in11} . \tag{4.1.1}$$

The stress tensor and its deviator are given as:

$$\boldsymbol{\sigma} = \begin{pmatrix} S & 0 & 0 \\ 0 & 0 & 0 \\ 0 & 0 & 0 \end{pmatrix} \quad , \quad \boldsymbol{\sigma}^* = \begin{pmatrix} \frac{2}{3}S & 0 & 0 \\ 0 & -\frac{1}{3}S & 0 \\ 0 & 0 & -\frac{1}{3}S \end{pmatrix} . \tag{4.1.2}$$

For the inelastic strain tensor $\boldsymbol{\varepsilon}_{in}$ as well as for the back stress $\boldsymbol{X}_{in}$ it holds that

$$\boldsymbol{\varepsilon}_{in} = \begin{pmatrix} \epsilon_{in} & 0 & 0 \\ 0 & -\frac{1}{2}\epsilon_{in} & 0 \\ 0 & 0 & -\frac{1}{2}\epsilon_{in} \end{pmatrix} \quad , \quad \boldsymbol{X}_{in} = \begin{pmatrix} X_{in11} & 0 & 0 \\ 0 & -\frac{1}{2}X_{in11} & 0 \\ 0 & 0 & -\frac{1}{2}X_{in11} \end{pmatrix} . \tag{4.1.3}$$

The norm of the inelastic strain tensor $\boldsymbol{\varepsilon}_{in}$ in (4.1.3) is

$$||\boldsymbol{\varepsilon}_{in}|| = \sqrt{\frac{3}{2}}|\epsilon_{in}| , \tag{4.1.4}$$

(see definition (2.2.11)). Hence, for the accumulated inelastic strain defined in (2.1.54), we get

$$\dot{s}_{in} = |\dot{\epsilon}_{in}| . \tag{4.1.5}$$

By means of the relations in (4.1.3), the von Mises equivalent stress in the uniaxial case is given by

$$\sqrt{\frac{3}{2}}||\boldsymbol{\sigma}^* - \boldsymbol{X}_{in}|| = |S - x_{in}| , \tag{4.1.6}$$

with the definition

$$x_{in} := \frac{3}{2}X_{in11} . \tag{4.1.7}$$

Finally, we are able to formulate the one-dimensional versions of the material laws for creep and TRIP, which will be presented in the following Subsection.

## 4.1.2 One-dimensional version of material laws

Using (4.1.6), the one-dimensional version of the material law for the creep strain (2.2.30) is:

$$\dot{\epsilon}_c = A \left( \frac{|S - x_c|}{D_c} \right)^{m-1} \frac{S - x_c}{D_c} s_c^k , \tag{4.1.8}$$

where $x_c$ is defined as in (4.1.7), i.e. $x_c := \frac{3}{2} X_{c11}$. The material parameters $A > 0$, $m > 0$ and $k$ generally depend on the temperature $\theta$ and possibly on further quantities. The uniaxial stress is denoted by $S$, $x_c$ and $D_c$ are called creep back stress and drag stress, respectively. Here, we assume a constant drag stress. In general, one may assume an additional evolution equation for $D_c$.

The one-dimensional form of the accumulated creep strain $s_c$ is defined by

$$s_c(t) = \int_0^t |\dot{\epsilon}_c(\tau)| \, \mathrm{d}\tau . \tag{4.1.9}$$

In the case of $x_c \equiv 0$, the material parameter $k$ controls the creep stage: $k < 0$ for primary creep, $k = 0$ for secondary creep and $k > 0$ considering tertiary creep. Here, we restrict ourselves to primary creep, thus $k < 0$ in the following. The evolution of the 1D creep back stress is described by

$$\dot{x}_c = c_c(\theta) \, \dot{\epsilon}_c - b_c(\theta, S) \, x_c \, (\dot{s}_c)^l + \frac{\mathrm{d}c_c}{\mathrm{d}\theta} \frac{\dot{\theta}}{c_c(\theta)} x_c , \tag{4.1.10}$$

where $c_c > 0$, $b_c \geq 0$ are material parameters generally depending on temperature (and possibly on stress) and $l \in \{0, 1\}$ (see remarks concerning equation (2.2.34) in Section 2.2.2).

The uniaxial variant of the material law for TRIP (2.2.39) is given by

$$\dot{\epsilon}_{trip} = \kappa \, S \, \frac{\mathrm{d}\phi}{\mathrm{d}p} \dot{p} , \tag{4.1.11}$$

where $\kappa > 0$ and $p$ denotes the fraction of the forming phase where $\dot{p} \geq 0$.

**Remark 4.1.1.** *In the case of TRIP with back stress, the evolution of the 1D back stress $x_{trip} := \frac{3}{2} X_{trip11}$ is given by (cf. Remark 2.2.3)*

$$\dot{x}_{trip} = c_{trip}(\theta) \, \dot{\epsilon}_{trip} - b_{trip}(\theta, S) \, x_{trip} \, \dot{s}_{trip} + \frac{\mathrm{d}c_{trip}}{\mathrm{d}\theta} \frac{\dot{\theta}}{c_{trip}(\theta)} x_{trip} . \tag{4.1.12}$$

In order to solve the equations (4.1.8), (4.1.10), the knowledge of the material parameters $A, m, k, c_c, b_c$ is presumed. Therefore, a parameter identification is performed. This requires appropriate experimental data. Section 4.2 describes the necessary experimental data and how to obtain further information from the data.

According to the equivalence hypothesis, the material parameters involved in the 3D equations and in the corresponding 1D version have the same meaning. As a consequence, after having determined the material parameters for (4.1.8), (4.1.10), we are able to handle the 3D problem using the obtained parameters.

## 4.2 Deriving further data from uniaxial experiments

The given quantities are: temperature $\theta$, length $l$ and diameter $d$ of the specimen, as well as applied stress $S$ as discrete functions of time $t$ (see Figure 4.1).

In the experiments considered here, stress and temperature are controlled to remain constant during the experiment (cf. Section 6 for details). Generally, the procedure presented in Section 4.3 is applicable for experiments under varying stress, temperature and strains. For further information, see Bökenheide and Wolff (2012), Wolff et al. (2012c), Wolff et al. (2012a), Wolff et al. (2011a).

Our aim is to obtain a maximum of information from given experimental data. For a detailed description of the approach, we refer to Wolff et al. (2012c). Further information about the performed experiments will be presented in Chapter 6. Details about experimental setup, technical details about the testing device and specimen geometry are given in Dalgic et al. (2009).

Based on Hooke's law and assuming spatial homogeneity we get the following equations for the (whole) longitudinal strain $\epsilon_L$ and the (whole) transversal strains $\epsilon_D$:

$$\epsilon_L(t) = \frac{l(t) - l_0}{l_0} = \frac{S(t)}{E(\theta(t), p(t))} + \sqrt[3]{\frac{\varrho_0}{\varrho(\theta(t), p(t))}} - 1 + \epsilon_{in} \quad = \varepsilon_{11}, \qquad (4.2.1)$$

$$\epsilon_D(t) = \frac{d(t) - d_0}{d_0} = \frac{-\nu(\theta(t), p(t))}{E(\theta(t), p(t))} S(t) + \sqrt[3]{\frac{\varrho_0}{\varrho(\theta(t), p(t))}} - 1 - \frac{1}{2}\epsilon_{in} \quad = \varepsilon_{22} = \varepsilon_{33}, \qquad (4.2.2)$$

where $E$ stands for Young's modulus, $\nu$ for Poisson's ratio, each depending on current temperature $\theta(t)$ and phase fractions $p(t) = (p_1(t), \dots, p_{M_p}(t))$. $\varrho$ stands for the density depending on $\theta(t)$ and $p(t)$, $\varrho_0$ denotes the density corresponding to the temperature and phase fractions at time $t = 0$, thus $\varrho_0 = \varrho(\theta_0, p_0)$. The values $l_0$ and $d_0$ are the initial length and diameter at the beginning of the experiment.

**Remark 4.2.1** (Linear coefficient of thermal expansion). *Based on* (2.1.52) *and* (4.1.3), *equations* (4.2.1) *and* (4.2.2) *are in accordance with* (2.1.49) *and with* (2.1.50). *In order to see this, the root in* (4.2.1) *and* (4.2.2) *can be linearised (in the case of no phase changes) by*

$$\sqrt[3]{\frac{\varrho_0}{\varrho(\theta(t))}} - 1 \approx \frac{\varrho_0 - \varrho(\theta(t))}{3\varrho_0} \approx -\frac{1}{3\varrho_0}\frac{\mathrm{d}\varrho}{\mathrm{d}\theta}(\theta_0)(\theta - \theta_0). \qquad (4.2.3)$$

*Thus, approximately we have*

$$\alpha_\theta = -\frac{1}{3\varrho_0}\frac{\mathrm{d}\varrho}{\mathrm{d}\theta}(\theta_0), \qquad (4.2.4)$$

*where $\alpha_\theta$ is the linear coefficient of thermal expansion in* (2.1.49).

Equations (4.2.1) and (4.2.2) imply important formulas for data processing:

- Volume strain:

$$\epsilon_V(t) = \mathrm{tr}(\boldsymbol{\varepsilon}) = \epsilon_L(t) + 2\,\epsilon_D(t)$$
$$= \frac{(1 - 2\,\nu(\theta(t), p(t)))}{E(\theta(t), p(t))}\, S(t) + 3\left(\sqrt[3]{\frac{\varrho_0}{\varrho(\theta(t), p(t))}} - 1\right) \tag{4.2.5}$$

- Difference of longitudinal and transversal strain:

$$\epsilon_L(t) - \epsilon_D(t) = \frac{(1 + \nu(\theta(t), p(t)))}{E(\theta(t), p(t))}\, S(t) + \frac{3}{2}\,\epsilon_{in}\ . \tag{4.2.6}$$

**Remark 4.2.2** (Volume strain). *In (4.2.5), we assume that the considered deformations are small, cf. Feynman et al. (1991).*

Note that the factors containing $E$ and $\nu$ in (4.2.5) and (4.2.6) can be expressed by means of the compression modulus $K$ and the shear modulus $\mu$ (cf. Remark 2.1.4):

$$\frac{1}{3\,K} = \frac{1 - 2\,\nu}{E}\quad,\qquad \frac{1}{2\,\mu} = \frac{1 + \nu}{E}\ . \tag{4.2.7}$$

By means of (4.2.6) and (4.2.7), we obtain a formula for the (total) inelastic longitudinal strain $\epsilon_{in}$ expressed via the (total) longitudinal strain $\epsilon_L$ and the (total) transversal strains $\epsilon_D$:

$$\epsilon_{in}(t) = \frac{2}{3}(\epsilon_L(t) - \epsilon_D(t)) - \frac{2(1 + \nu(\theta(t), p(t)))}{3E(\theta(t), p(t))}\, S(t) \tag{4.2.8}$$
$$= \frac{2}{3}(\epsilon_L(t) - \epsilon_D(t)) - \frac{S(t)}{3\mu(\theta(t), p(t))}\ . \tag{4.2.9}$$

This formula enables us to obtain the (total) inelastic longitudinal strain by means of the given measured data.

For further remarks and special cases, we refer to Wolff et al. (2012c).

**Remark 4.2.3.**

(i) *A nice feature of formulas (4.2.5) and (4.2.6) is that isotropic and anisotropic effects are separated. Equation (4.2.5) does not contain the inelastic strain $\epsilon_{in}$. Knowing the dependence of $K$ and $\varrho$ on temperature $\theta$ and phase fractions $p$ (and exploiting mixture rules and the phase balance), we can calculate the evolution of the forming phase without assuming any special law of phase transformation. See Wolff et al. (2012c) and Wolff et al. (2012a) for details.*

(ii) *The density term does not occur in (4.2.9). Assuming that $\mu$ does not depend on the phase fractions $p$, the right-hand side of formula (4.2.9) is independent of possible phase transformations. Frequently the elastic part is neglected, especially for a constant stress $S$ during phase changes (cf. Ahrens et al. (2000), Dalgic and Löwisch (2006), e.g.). Otherwise, $p$ must be determined via (4.2.5).*

*(iii) Alternatively to (4.2.9), we can derive a formula for the inelastic strain $\epsilon_{in}$ using (4.2.1):*

$$\epsilon_{in}(t) = \epsilon_L(t) - \frac{S(t)}{E(\theta(t), p(t))} - \left( \sqrt[3]{\frac{\varrho_0}{\varrho(\theta(t), p(t))}} - 1 \right) . \qquad (4.2.10)$$

*In the case of no phase transformations (e.g. in cyclic plasticity), formula (4.2.10) is frequently used.*

*If there is no transversal strain available in the given experimental data, formula (4.2.10) can be applied to determine the inelastic strain.*

## 4.3  Verification of material laws using uniaxial experimental data

By means of the evaluation of experimental data and the observed material response, we can assume an underlying specific material law (e.g. primary creep).

Therefore, the knowledge of certain material parameters is required. Our aim is to determine these parameters in such a way that the simulation results give us the best possible fit to the considered experimental data set.

We use the formulas provided in Section 4.2, in order to derive further data from the experimental data. This enables us to perform a parameter identification.

The identification of the parameters, i.e. the *'inverse problem'* will be handled in Section 4.6. First we have to deal with the forward problem - the *'direct problem'*. This means to solve the discretised model equations assuming specific material laws and parameters to be given. For details about the numerical algorithm we refer to Wolff et al. (2012c).

In uniaxial experiments, the given data usually consists of the longitudinal strain, the transversal strain and the stress as discrete functions of time (see Section 4.2). Considering non-isothermal experiments, the temperature and - in case of phase transformation - the fractions of the considered phases are given, too.

We will present two different approaches in order to deal with the direct problem: the so-called *'strain-driven'* and the *'stress-driven'* approach. In these approaches we assume a part of the experimental data as given and have to calculate the remaining unknowns.

First, we present both approaches for general inelastic material behaviour. Therefore, we consider the discretised version of the one-dimensional equations presented in Section 4.1.2. Section 4.4 will provide the proceeding in the case of creep behaviour, Section 4.5 in the case of creep and TRIP.

We briefly sketch both approaches and give some information about parameter identification before we go into further detail.

**Direct problem: Strain-driven approach**  The longitudinal and transversal strains are given as discrete functions of time. Sometimes, only the transversal strain is given. In the non-isothermal case, the temperature and possibly the phase fractions are given, too.

Assuming a specific material behaviour (i.e. the knowledge of material laws and parameters), we aim to calculate the material response, i.e. stress, inelastic strains, back stresses etc. Thus, the strain-driven approach allows the simulation of strain-controlled experiments. In Subsection 4.3.1, we provide further details. In Subsections 4.4.2 and 4.5.2, we apply the approach to creep and to TRIP, respectively.

**Direct problem: Stress-driven approach** Now, the uniaxial stress is given as a discrete function of time. Moreover, in the non-isothermal case, the temperature and possibly the phase fractions are given, too. Again, our aim is to calculate the inelastic strain. In order to simulate the total longitudinal (and transversal) strain, we need additional information, e.g. concerning the density as a function of temperature and phase fractions. The stress-driven approach allows the simulation of stress-controlled experiments (e.g. in ratcheting).

In Subsection 4.3.2 we provide further details about the approach. In Subsections 4.4.3 and 4.5.3, we apply the stress-driven approach to creep and TRIP, respectively.

**Inverse problem: Identification of material parameters** The evaluation of experimental data gives us information about the material response. Therefore, we can assume a specific material law. Our aim is to determine the material parameters of the underlying model in such a way that the simulation results give us the best possible fit to the considered experimental data set. In Section 4.6, we provide the general strategy as well as the application to concrete material laws.

As pointed out in Mahnken and Stein (1996), in the case of complex material behaviour an optimisation procedure which allows a *simultaneous* determination of several parameters is required. We will follow this approach, see Section 4.6.

## 4.3.1 Direct problem: General strain-driven approach

At the beginning of the $n^{th}$ time step, only the longitudinal and transversal strains $\epsilon_L^n$ and $\epsilon_D^n$ as well as possibly the temperature $\theta^n$ and phase fractions $p^n$ are given. The corresponding stress $S^n$ and inelastic strain $\epsilon_{in}^n$ have to be calculated using the equations governing the material behaviour. In many cases, the material law for inelastic behaviour has the general rate form

$$\dot{\epsilon}_{in} = g(S, x, \dots), \qquad (4.3.1)$$

where $g$ stands for the material law. The laws for creep or TRIP given in (4.1.8) and (4.1.11), respectively, have this rate form. Using (4.2.9), we obtain an equation for the discrete value of the stress $S^n$:

$$S^n = \frac{E_n}{1 + \nu_n} \left( \epsilon_L^n - \epsilon_D^n - \frac{3}{2} \epsilon_{in}^n \right), \qquad (4.3.2)$$

where we use the abbreviations $E_n = E(\theta^n, p^n)$, $\nu_n = \nu(\theta^n, p^n)$ etc. for the material parameters. Rearranging (4.3.2), we obtain

$$S^n = \frac{E_n}{1 + \nu_n} \left( \epsilon_L^n - \epsilon_D^n - \frac{3}{2} \epsilon_{in}^{n-1} \right) - \frac{3E_n}{2(1 + \nu_n)} \left( \epsilon_{in}^n - \epsilon_{in}^{n-1} \right). \qquad (4.3.3)$$

We introduce the trial stress for the uniaxial case:

$$S_{trial}^n := \frac{E_n}{1 + \nu_n} \left( \epsilon_L^n - \epsilon_D^n - \frac{3}{2} \epsilon_{in}^{n-1} \right). \qquad (4.3.4)$$

The current value of the inelastic strain follows by

$$\epsilon_{in}^n = \epsilon_{in}^{n-1} + \tau_n \left( \dot{\epsilon}_{in} \right)^n, \qquad (4.3.5)$$

(see (5.2.2)) where $\tau_n := t_n - t_{n-1}$. The quantity $(\dot{\epsilon}_{in})^n$ stands for a suitable discretisation of the time derivative of the inelastic strain.

Equations (4.3.3)–(4.3.5), yield the following approximation for $S^n$:

$$S^n = S_{trial}^n - \frac{3E_n}{2(1 + \nu_n)} \tau_n(\dot{\epsilon}_{in})^n. \qquad (4.3.6)$$

In non-isothermal cases, the approach in (4.3.6) has the advantage that it does not contain density terms. Using (4.3.1), we can set

$$(\dot{\epsilon}_{in})^n = g(S^n, x^n, \dots). \qquad (4.3.7)$$

Inserting (4.3.6) in (4.3.7) shows that the required value $(\dot{\epsilon}_{in})^n$ is only implicitly given. In order to deal with this problem, we have to find an appropriate solution scheme. Additionally, an implicit approach for the back stress $x^n$ must be added. After calculations and updates, we obtain the discrete values for the stress $S^n$ as well as for the inelastic strain $\epsilon_{in}^n$. In Sections 4.4 and 4.5, we will describe the procedure for the case of creep as well as for TRIP, respectively.

**Remark 4.3.1.** (i) The stress $S^n$ given by (4.3.6) is usually called 'corrected stress'. in the case of plasticity or viscoplasticity, the material law involves a flow rule that is used to decide whether the current time step lies in the elastic domain. In a 'plastic' time step, the plastic multiplier has to be determined simultaneously. This leads to the so-called predictor-corrector approach. We refer to Simo and Hughes (1998) for details.

(ii) The authors in Mahnken and Stein (1996) apply a similar approach to viscoplasticity using different techniques (midpoint rule, Newton's method).

(iii) *The presented approach is also applicable in the case of several inelastic phenomena. In this case, $\epsilon_{in}$ is a sum of several parts. For each of them the material law must be discretised according to (4.3.7).*

(iv) *If no transversal strain is given in the experimental data (cf. Remark 4.2.3), instead of formulas (4.3.4) and (4.3.6) we set*

$$S_{trial}^n := E_n \left( \epsilon_L^n - \left( \sqrt[3]{\frac{\varrho_0}{\varrho(\theta^n, p^n)}} - 1 \right) - \epsilon_{in}^{n-1} \right), \qquad (4.3.8)$$

$$S^n := S_{trial}^n - E_n \, \tau_n(\dot{\epsilon}_{in})^n \,. \qquad (4.3.9)$$

### 4.3.2 Direct problem: General stress-driven approach

This approach assumes that in the $n^{th}$ time step, the stress $S^n$ and possibly the temperature $\theta^n$ and phase fraction $p^n$ are known whereas the strains $\epsilon_L^n$, $\epsilon_D^n$ and $\epsilon_{in}^n$ (and possibly further inelastic quantities) must be calculated.

In the same manner as in the strain-driven approach we use the approximation

$$\epsilon_{in}^n = \epsilon_{in}^{n-1} + \tau_n(\dot{\epsilon}_{in})^n \,, \qquad (4.3.10)$$

with the discretisation

$$(\dot{\epsilon}_{in})^n = g(S^n, x^n, \dots) \,. \qquad (4.3.11)$$

After further discretisations of the back stress $x$ and other quantities, we obtain an implicit equation for $(\dot{\epsilon}_{in})^n$ from (4.3.11). Having solved this equation, we can update the inelastic strain $\epsilon_{in}^n$ by (4.3.10). In the next step, we calculate the difference between the longitudinal strain $\epsilon_L^n$ and the transversal strain $\epsilon_D^n$ by

$$\epsilon_L^n - \epsilon_D^n = \frac{1 + \nu_n}{E_n} S^n + \frac{3}{2} \epsilon_{in}^n \,. \qquad (4.3.12)$$

using equation (4.2.6).

In the non-isothermal case, the calculation of the longitudinal strain $\epsilon_L^n$ requires additional information. If the density is given as a function of temperature and phase fractions, we have

$$\epsilon_L^n = \frac{1}{E_n} S^n + \left( \sqrt[3]{\frac{\varrho_0}{\varrho(\theta^n, p^n)}} - 1 \right) + \epsilon_{in}^n \,. \qquad (4.3.13)$$

After that, $\epsilon_D^n$ can be calculated by

$$\epsilon_D^n = -\frac{\nu_n}{E_n} S^n + \left( \sqrt[3]{\frac{\varrho_0}{\varrho(\theta^n, p^n)}} - 1 \right) - \frac{1}{2} \epsilon_{in}^n \,. \qquad (4.3.14)$$

In Sections 4.4.3 and 4.5.3, we will apply the approach method to creep and TRIP material behaviour, respectively.

## 4.4 Verification of material laws for creep

Now, we will apply the approaches presented in Section 4.3.1 and 4.3.2 to the equations in (4.1.8), (4.1.10).

### 4.4.1 Discretised version of the model equations (1D)

We follow Wolff et al. (2012c), choosing a fully implicit approach for the one-dimensional material law for creep (4.1.8),

$$(\dot{\epsilon}_c)^n = A(\theta_n) \left( \frac{|S^n - x_c^n|}{D_c} \right)^{m(\theta_n)-1} \frac{S^n - x_c^n}{D_c} \, (s_c^n)^{k(\theta_n)} , \qquad (4.4.1)$$

and a semi-implicit approach for the back stress (4.1.10):

$$(\dot{x}_c)^n = c_c(\theta^n) \, (\dot{\epsilon}_c)^n - b_c(\theta^n, S^{n-1}) \, x_c^n \, ((\dot{s}_c)^{n-1})^l + \left( \frac{dc_c}{d\theta} \right)^n \frac{(\dot{\theta})^n}{c_c(\theta^n)} \, x_c^{n-1} . \qquad (4.4.2)$$

Furthermore, from (4.1.9) with the approximation in (5.2.2), we obtain

$$s_c^n = s_c^{n-1} + \tau_n |(\dot{\epsilon}_c)^n| , \qquad (4.4.3)$$

where $\tau_n = t_n - t_{n-1}$.

The material parameters in (4.4.1)–(4.4.2), i.e. $A_n$, $m_n$, $k_n$, $c_{c,n}$, $b_{c,n}$ are determined via a parameter identification. The procedure will be presented in detail in Section 4.6. Here, we assume the parameters to be given.

### 4.4.2 Direct problem: Strain-driven approach for creep

We apply the approach presented in Section 4.3.1 to creep material behaviour, thus $\epsilon_{in} = \epsilon_c$. In the strain-driven approach, the values of the strains $\epsilon_L^n$ and $\epsilon_D^n$, as well as temperature $\theta^n$ and phase fractions $p^n$ are assumed to be known. Our aim is to calculate the value of the stress $S^n$ as well as the value of the current creep strain $\epsilon_c^n$ and the back stress $x_c^n$.

First, we introduce the effective stress $\xi^n$ by the definition

$$\xi^n := S^n - x_c^n . \qquad (4.4.4)$$

In the following, we will solve for the unknowns in the following order:

$$(\dot{\epsilon}_c)^n , \; \xi^n , \; \epsilon_c^n , \; s_c^n , \; S^n , \; x_c^n . \qquad (4.4.5)$$

Using the equation for the stress $S^n$ in (4.3.6) and the definition of the trial stress in (4.3.4), we obtain for the effective stress

$$\xi^n = S_{trial}^n - x_c^n - \frac{3E_n}{2(1 + \nu_n)} \tau_n (\dot{\epsilon}_c)^n. \qquad (4.4.6)$$

By means of the discretised evolution equation (4.4.2) of the back stress $x_c^n$, we obtain the following approach for the back stress $x_c^n$:

$$x_c^n = x_c^{n-1} + \tau_n c_{c,n} \left(\dot{\epsilon}_c\right)^n - \tau_n b_{c,n} x_c^n \left((\dot{s}_c)^{n-1}\right)^l + \frac{c_{c,n} - c_{c,n-1}}{c_{c,n}} x_c^{n-1} , \qquad (4.4.7)$$

where $c_{c,n} = c_c(\theta^n)$, $b_{c,n} = b_c(\theta^n, S^{n-1})$. Note that $l$ equals 0 or 1. Using the evolution equation for the creep strain (4.4.1), the definition of the effective stress (4.4.4) and the discretisation of the accumulated creep strain $s_c^n$

$$s_c^n = s_c^{n-1} + \tau_n \left|(\dot{\epsilon}_c)^n\right| , \qquad (4.4.8)$$

we obtain

$$(\dot{\epsilon}_c)^n = A_n \left(\frac{|\xi^n|}{D_c}\right)^{m_n - 1} \frac{\xi^n}{D_c} \left(s_c^{n-1} + \tau_n \left|(\dot{\epsilon}_c)^n\right|\right)^{k_n} , \qquad (4.4.9)$$

with $A_n := A(\theta^n)$, $m_n := m(\theta^n)$, $k_n := k(\theta^n)$ and a constant drag stress $D_c$. Solving equation (4.4.7) with respect to $x_c^n$ yields

$$x_c^n = \frac{x_c^{n-1} + \tau_n c_{c,n} \left(\dot{\epsilon}_c\right)^n + \frac{c_{c,n} - c_{c,n-1}}{c_{c,n}} x_c^{n-1}}{1 + \tau_n b_{c,n} \left(\left|(\dot{\epsilon}_c)^{n-1}\right|\right)^l} . \qquad (4.4.10)$$

Inserting (4.4.10) into (4.4.6), we obtain

$$\xi^n = S_{trial}^n - \frac{1 + \frac{c_{c,n} - c_{c,n-1}}{c_{c,n}}}{1 + \tau_n b_{c,n} \left(\left|(\dot{\epsilon}_c)^{n-1}\right|\right)^l} x_c^{n-1} +$$

$$- \left(\frac{3E_n}{2(1 + \nu_n)} + \frac{c_{c,n}}{1 + \tau_n b_{c,n} \left(\left|(\dot{\epsilon}_c)^{n-1}\right|\right)^l}\right) \tau_n (\dot{\epsilon}_c)^n . \qquad (4.4.11)$$

Equations (4.4.9) and (4.4.11) form a system which can be solved for $\xi^n$ and $(\dot{\epsilon}_c)^n$. Inserting (4.4.11) into (4.4.9), we obtain a non-linear equation for $(\dot{\epsilon}_c)^n$ that enables us to determine the current value of the time derivative $(\dot{\epsilon}_c)^n$.

After this, we update the remaining values (see (4.4.5)). We use (4.4.11) to calculate the effective stress $\xi^n$. The current value of the creep strain follows by

$$\epsilon_c^n = \epsilon_c^{n-1} + \tau_n (\dot{\epsilon}_c)^n , \qquad (4.4.12)$$

the accumulated creep strain $s_c^n$ by (4.4.8).

Using

$$S^n = \frac{E_n}{1 + \nu_n} \left(\epsilon_L^n - \epsilon_D^n - \frac{3}{2}\epsilon_c^n\right) , \qquad (4.4.13)$$

we obtain the current value of the stress. Finally, the back stress $x_c^n$ is determined by (4.4.10).

**Remark 4.4.1.** *(i) If the creep law is modelled without a back stress, the algorithm simplifies: Equation (4.4.9) reduces to*

$$(\dot{\epsilon}_c)^n = A_n \left(\frac{|S^n|}{D_c}\right)^{m_n-1} \frac{S^n}{D_c} \left(s_c^{n-1} + \tau_n |(\dot{\epsilon}_c)^n|\right)^{k_n} . \qquad (4.4.14)$$

*Inserting (4.3.6) into (4.4.14), we obtain an equation for $(\dot{\epsilon}_c)^n$ which can be (numerically) solved.*

*(ii) If the drag stress $D_c$ is a function, we have an additional evolution equation, see (2.2.35). It has to be discretised in a suitable way.*

*(iii) If only longitudinal strains are given, the values of $S_{trial}^n$ and $S^n$ must be calculated alternatively by (4.3.8) and (4.3.9), respectively.*

## 4.4.3 Direct problem: Stress-driven approach for creep

Now, in the $n^{th}$ time step, the stress $S^n$ and possibly $\theta^n$ and $p^n$ are assumed to be known. The strains $\epsilon_L^n$, $\epsilon_D^n$ as well as the creep strain $\epsilon_c^n$ must be calculated. In accordance with (4.4.1) and (4.4.3), the discretised creep-strain rate is given by

$$(\dot{\epsilon}_c)^n = A_n \left(\frac{|S^n - x_c^n|}{D_c}\right)^{m_n-1} \frac{S^n - x_c^n}{D_c} \left(s_c^{n-1} + \tau_n |(\dot{\epsilon}_c)^n|\right)^{k_n}, \qquad (4.4.15)$$

where now the stress $S^n$ is known. We use the discretisation of the back stress $x_c$ as specified in (4.4.7) which yields (4.4.10). Thus, the current value of the back stress is given by

$$x_c^n = \frac{x_c^{n-1} + \tau_n c_{c,n} (\dot{\epsilon}_c)^n + \frac{c_{c,n} - c_{c,n-1}}{c_{c,n}} x_c^{n-1}}{1 + \tau_n b_{c,n} \left(|(\dot{\epsilon}_c)^{n-1}|\right)^l} . \qquad (4.4.16)$$

In the next step, we insert (4.4.16) into (4.4.15) and obtain a non-linear equation for $(\dot{\epsilon}_c)^n$.

After having determined the value of $(\dot{\epsilon}_c)^n$, we can update the other quantities, i.e.:

$$\epsilon_c^n , \ s_c^n , \ \epsilon_L^n , \ \epsilon_D^n . \qquad (4.4.17)$$

using (4.4.12) and (4.4.8). The longitudinal and transversal strains follow by (4.3.13) and (4.3.14).

We calculate the difference

$$\epsilon_L^n - \epsilon_D^n = \frac{1 + \nu_n}{E_n} S^n + \frac{3}{2}\epsilon_c^n . \qquad (4.4.18)$$

Finally, the longitudinal and transversal strains are determined by

$$\epsilon_L^n = \frac{1}{E_n} S^n + \left(\sqrt[3]{\frac{\varrho_0}{\varrho(\theta^n, p^n)}} - 1\right) + \epsilon_c^n , \qquad (4.4.19)$$

and

$$\epsilon_D^n = -\frac{\nu_n}{E_n} S^n + \left( \sqrt[3]{\frac{\varrho_0}{\varrho(\theta^n, p^n)}} - 1 \right) - \frac{1}{2} \epsilon_c^n . \tag{4.4.20}$$

**Remark 4.4.2.** *1. If the creep law contains no back stress, equation (4.4.15) can be directly solved for* $(\dot{\epsilon}_c)^n$.

# 4.5 Verification of material laws for TRIP

In this section, we consider the inelastic strain as the sum of the TRIP and the creep strain, i.e. $\epsilon_{in} = \epsilon_c + \epsilon_{trip}$.

In the case of the simultaneous appearance of creep and TRIP, first we determine the current value of the creep strain. We use the respective approach (strain- or stress-driven) presented in Section 4.4.

## 4.5.1 Discretised version of the model equations (1D)

In the case of creep and TRIP, we have an additional evolution equation for the TRIP strain $\epsilon_{trip}$, see (4.1.11). The discretised evolution equation is

$$(\dot{\epsilon}_{trip})^n = \kappa_n S^n \frac{\mathrm{d}\phi}{\mathrm{d}p}(p^n) \frac{1}{\tau_n} (p^n - p^{n-1}), \tag{4.5.1}$$

where $\kappa_n = \kappa(\theta^n, S^{n-1})$. We assume $p^n \geq p^{n-1}$. In the following, we apply the schemes outlined in Subsections 4.3.1 and 4.3.2 to TRIP material behaviour. In the case of TRIP, the approaches are much easier than in the case of creep as the evolution of the TRIP strain is modelled without back stress or accumulated (TRIP) strain. The parameter $\kappa$ depends on temperature and stress. We use the value of the stress of the former time step, thus $\kappa_n = \kappa(\theta^n, S^{n-1})$. As already assumed above, the discrete values $p^n$ of the forming phase are supposed to be given at this step of the algorithm.

**Remark 4.5.1.** *(i) If we model the evolution of the TRIP strain with a back stress, the equation for* $x_{trip}$ *(4.1.12) has to be discretised similarly to the creep back stress* $x_c$ *in (4.4.7).*

*(ii) If there is only TRIP (as in the case of quenching, e.g.), we can apply the schemes which will be outlined in the following subsections by setting the creep strain values* $\epsilon_c^n$ *equal to zero.*

## 4.5.2 Strain-driven approach for TRIP

In this case, $\epsilon_L^n$, $\epsilon_D^n$, $\theta^n$, $p^n$ as well as $\epsilon_c^n$ are assumed to be known. Our aim is to determine the current value of the stress $S^n$ as well as of the TRIP strain $\epsilon_{trip}^n$.

First, we calculate the trial stress according to (4.3.4). We use the value of the creep strain calculated beforehand, thus we set $\epsilon_{in}^{n-1} = \epsilon_c^n + \epsilon_{trip}^{n-1}$:

$$S_{trial}^n = \frac{E_n}{1 + \nu_n} \left( \epsilon_L^n - \epsilon_D^n - \frac{3}{2} \epsilon_c^n - \frac{3}{2} \epsilon_{trip}^{n-1} \right) . \tag{4.5.2}$$

Due to equation (4.3.2) for the stress $S^n$, i.e.

$$S^n = \frac{E_n}{1 + \nu_n} \left( \epsilon_L^n - \epsilon_D^n - \frac{3}{2} (\epsilon_c^n + \epsilon_{trip}^n) \right) , \tag{4.5.3}$$

we obtain

$$S^n = S_{trial}^n - \frac{3E_n}{2(1 + \nu_n)} \tau_n (\dot\epsilon_{trip})^n , \tag{4.5.4}$$

with the definition of the trial stress introduced in (4.5.2). Inserting (4.5.1) into (4.5.4), leads to the equation

$$S^n = S_{trial}^n \left( \left( 1 + \frac{3E_n}{2(1 + \nu_n)} \right) \kappa_n \frac{d\phi}{dp}(p^n) \left( p^n - p^{n-1} \right) \right)^{-1} . \tag{4.5.5}$$

We insert the definition of $S_{trial}^n$ and obtain an equation which can be solved for $S^n$:

$$S^n = \frac{\frac{E_n}{1+\nu_n}(\epsilon_L^n - \epsilon_D^n - \frac{3}{2}\epsilon_c^n - \frac{3}{2}\epsilon_{trip}^{n-1})}{\left( 1 + \frac{3E_n}{2(1+\nu_n)} \right) \kappa_n \frac{d\phi}{dp}(p^n) \left( p^n - p^{n-1} \right)} . \tag{4.5.6}$$

After having calculated the stress $S^n$, we update the TRIP strain by

$$\epsilon_{trip}^n = \epsilon_{trip}^{n-1} + \kappa_n S^n \frac{d\phi}{dp} \left( p^n \right) \left( p^n - p^{n-1} \right) . \tag{4.5.7}$$

### 4.5.3 Stress-driven approach for TRIP

We follow the general approach presented in Section 4.3.2. We assume that the current value of the creep strain $\epsilon_c^n$ was determined beforehand. Thus, $\epsilon_c^n$ as well as $S^n$, $\theta^n$ and $p^n$ are known. Our aim is to calculate the difference $\epsilon_L^n - \epsilon_D^n$ and the strains $\epsilon_{trip}^n$, $\epsilon_L^n$, $\epsilon_D^n$. Since the stress $S^n$ is assumed to be known, the procedure is very simple:

The current value of the time derivative of the TRIP strain $(\dot\epsilon_{trip})^n$ can be calculated directly by the discretised material law (4.5.1),

$$(\dot\epsilon_{trip})^n = \kappa_n S^n \frac{d\phi}{dp}(p^n) \frac{1}{\tau_n} \left( p^n - p^{n-1} \right) , \tag{4.5.8}$$

which gives us the current trip strain:

$$\epsilon_{trip}^n = \epsilon_{trip}^{n-1} + \tau_n (\dot\epsilon_{trip})^n . \tag{4.5.9}$$

The difference between $\epsilon_L^n$ and $\epsilon_D^n$ can now be calculated by

$$\epsilon_L^n - \epsilon_D^n = \frac{1 + \nu_n}{E_n} S^n + \frac{3}{2} \epsilon_c^n + \frac{3}{2} \epsilon_{trip}^n \,. \tag{4.5.10}$$

Finally, the longitudinal and transversal strains are determined by

$$\epsilon_L^n = \frac{1}{E_n} S^n + \left( \sqrt[3]{\frac{\varrho_0}{\varrho(\theta^n, p^n)}} - 1 \right) + \epsilon_c^n + \epsilon_{trip}^n \,, \tag{4.5.11}$$

and

$$\epsilon_D^n = -\frac{\nu_n}{E_n} S^n + \left( \sqrt[3]{\frac{\varrho_0}{\varrho(\theta^n, p^n)}} - 1 \right) - \frac{1}{2} \epsilon_c^n - \frac{1}{2} \epsilon_{trip}^n \,. \tag{4.5.12}$$

The material parameters $\kappa_n$ and the saturation function $\phi_n$ for the TRIP strain are determined via a parameter identification. The procedure will be the same as in the case of creep (cf. Section 4.6.4). We will present the approach for the latter case in detail in Section 4.6.

## 4.6 Parameter identification

In the previous Sections 4.4–4.5, we presented the 1D model equations, their discretisation as well as the numerical solution approach. In order to perform simulations, the knowledge of the material parameters in the corresponding material laws is required.

Generally, our aim is to simulate the spatially homogeneous, uniaxial material behaviour. By means of experimental data we are able to identify the corresponding material behaviour and to assume a specific material law. In order to determine certain material parameters required for the underlying material laws, we conduct a parameter identification taking experimental data and calculated values into account. Here, our aim is to reduce the deviation between experimental and simulation results in order to obtain a best possible fit to the given data.

We refer to Wolff et al. (2012c), Bökenheide et al. (2011), Bökenheide et al. (2012b) and Bökenheide and Wolff (2012) for further details. For a further discussion about general aspects of parameter identification see Mahnken and Stein (1996), Mahnken (2004) and the references therein.

In Yun and Shang (2011), a 3D optimisation procedure was developed which uses forces and displacements on the same partial boundaries. That approach generalises the uniaxial homogeneous setting.

### 4.6.1 Experimental data

We consider experiments as described in Section 4.2. The following data is given by a performed uniaxial experiment: current length $l$ and diameter $d$ of the specimen, stress $S$ and temperature $\theta$. The data is given as discrete functions of time $t$.

As demonstrated in Section 4.2, we are able to obtain further information by the experimental data. The longitudinal and transversal strains as well as the inelastic strain can be derived from the given data using (4.2.1), (4.2.2) and (4.2.9) and can thus be assumed as given, too.

Altogether, the following (experimental) data is assumed to be given:

- $\epsilon_{L,exp}$ , $\epsilon_{D,exp}$ , $\epsilon_{in,exp}$ , $S_{exp}$ , $\theta_{exp}$ .

In the experimental results presented in Chapter 6, we will consider a specific set of experiments (i.e. for a specific phase). Each was performed under a different (constant) temperature. Our aim is to obtain the material parameters which represent the best fit taking *all* of the performed experiments into account. Therefore, we set each parameter as a parameter function depending (linearly) on the temperature $\theta$.

## 4.6.2 General strategy

We use an optimisation procedure which allows the simultaneous determination of the required parameters. We summarise the required material parameters in the set $\wp$. In order to find the optimal parameter set, we consider a cost functional and minimise it with respect to $\wp$.

Let $\mathcal{P}$ denote the collection of all possible parameter sets over which we vary our parameters in order to reduce the error between measured and calculated values.

We assume that a number of $N_{exp}$ experiments is given and set the number of data points of a test $j \in \{1, \ldots, N_{exp}\}$ as $N_j$. Let $\Phi$ define a cost functional (e.g. $l^2$ norm) representing the error between the values obtained by the experiment and the values predicted by the model. The considered cost functional can take for instance the values of inelastic strain, stress, etc. into account. Furthermore, it is also possible to consider only one test of an experimental set individually and to determine the optimal parameters for this test.

Our aim is to minimise the sum over all $N_{exp}$ tests of the error between the measured and the calculated values:

$$\min_{\wp \in \mathcal{P}} \sum_{j=1}^{N_{exp}} \Phi(\epsilon_{in,exp}^j, S_{exp}^j, \epsilon_{in,cal}^j, S_{cal}^j, \wp) . \tag{4.6.1}$$

The inelastic strains $\epsilon_{in,cal}$ can be calculated using one of the approaches presented in Section 4.3. Regarding the cost functional, it is also possible to include both the results of the strain-driven and the stress-driven approach (cf. Section 4.6.3).

We summarise the general optimisation strategy in Box 4.6.1.

**4.6.1. General optimisation strategy**

- Pre-process
    - Given: experimental data, i.e. samples of the form
      $\{t\ ,\ l\ ,\ d\ ,\ S\ ,\ \theta\}$.
    - Obtain further values $\quad\epsilon_{L,exp}$ , $\epsilon_{D,exp}$ , $\epsilon_{in,exp}$.
    - Set possible constraints for the parameters, all possible
      combinations of parameters form the set $\mathcal{P}$
    - Choose initial parameter set $\wp_0 \in \mathcal{P}$

- Optimisation:
    - Find optimal parameter set $\hat{\wp} \in \mathcal{P}$ that minimises the cost
      functional $\Phi(\epsilon_{in,exp}, S_{exp}, \epsilon_{in,cal}, S_{cal}, \wp)$

In Section 4.6.3, we will focus on the parameter identification in the case of creep. We will present the identification of the parameters required in the creep material law (4.4.1),(4.4.2). Considering the cost functional in (4.6.1), we will use the $l^2$ norm, see Section 4.6.3. The applied experimental data, the results of the parameter identification as well as the resulting creep strain in comparison with the experimental data will be presented in detail in Chapter 6.

In Section 4.6.3 and 4.6.4, we will handle the optimisation strategy in the case of creep and trip, respectively.

## 4.6.3 Parameter identification in the case of creep

In order to describe creep behaviour, we consider the discretised form of the material laws (4.4.1),(4.4.2) presented in Section 4.4. Our aim is to determine the required material parameters by finding the best approximation of the material behaviour compared to experimental data. Here, it is possible to either refer to one experiment or to all tests of a specific set of performed experiments (e.g. for a specific phase). Chapter 6 will describe the performed experiments in detail. The experiments are performed under different fixed temperatures and stresses.

Three different approaches for the cost functional in (4.6.1) are possible: Taking one approach – i.e. either the strain-driven or the stress-driven approach – or taking both approaches into account. See Sections 4.4.2 and 4.4.3 for details about the procedures.

In the following, we specify the approach outlined in Section 4.6.2 for creep material behaviour. We consider the model (4.4.1),(4.4.2). Our aim is to determine the parameter set

$$\wp_c = \{A, m, k, c_c, b_c\} \tag{4.6.2}$$

which yields the best fit to given experimental results.

We assume a number of $N_{exp}$ tests to be given where the number of data points in the $j$-th experiment is $N_j$, $j = 1, \ldots, N_{exp}$. By taking all tests of an experimental set (performed under different temperatures) into account, we are able to identify parameter functions $\wp_{c,\theta}$ depending on temperature.

Using (4.2.1), (4.2.2), we can assume the following data to be given for each creep experiment $j$:

$$\left[ (\epsilon_{L,exp})^i \, , \, (\epsilon_{D,exp})^i \, , \, (\epsilon_{c,exp})^i \, , \, (S_{exp})^i \, , \, (\theta_{exp})^i \right]_j \tag{4.6.3}$$

where $j \in \{1, \ldots, N_{exp}\}$ and $i = 1, \ldots, N_j$. The experimental creep strain $\epsilon_{c,exp}$ can easily be calculated by means of (4.2.9) with $\epsilon_{in} = \epsilon_c$ and can thus be considered as given, too.

After that, suitable sets of parameters $\mathcal{P}_c$ and start values $\wp_{c0} \in \mathcal{P}_c$ have to be chosen. Generally, our aim is to find the parameter set which minimizes

$$\min_{\wp \in \mathcal{P}_c} \sum_{j=1}^{N_{exp}} \Phi(\epsilon_{c,exp}^j, S_{exp}^j, \epsilon_{c,cal}^j, S_{cal}^j, \wp) \,, \tag{4.6.4}$$

where the cost functional $\Phi$ has to be chosen appropriately. Here, we use a least square approach of the difference of the measured and the calculated values. Hence, taking both the strain-driven *and* the stress-driven approach into account, we have to determine the parameter set $\{A, m, k, c_c, b_c\}$ with respect to

$$\min_{\wp \in \mathcal{P}_c} \sum_{j=1}^{N_{exp}} \left[ \left( \sum_{i=1}^{N_j} \left| \epsilon_{c,exp}^{n,j} - \epsilon_{c,cal\_strain}^{n,j} \right|^2 \right)^{1/2} + \left( \sum_{i=1}^{N_j} \left| \epsilon_{c,exp}^{n,j} - \epsilon_{c,cal\_stress}^{n,j} \right|^2 \right)^{1/2} \right] \,. \tag{4.6.5}$$

The values $\epsilon_{c,cal\_strain}^{i,j}$, $\epsilon_{c,cal\_stress}^{i,j}$ are the values of the calculated creep strain using the strain-driven and the stress-driven approach, respectively. The experimental creep strain is denoted by $\epsilon_{c,exp}^{i,j}$.

Taking several experiments under different temperatures into account, we are able to identify the coefficients of parameter functions depending on temperature. In order to obtain temperature depending parameters we set:

$$A(\theta) = A_0 \exp\left( \frac{-Q}{R\,\bar{\theta}} \right) \,, \qquad m(\theta) = m_0 + m_1\theta, \qquad k(\theta) = k_0 + k_1\theta \,, \tag{4.6.6}$$

$$c_c(\theta) = c_0 + c_1\theta, \qquad b_c(\theta) = b_0 + b_1\theta,$$

(cf. Chapter 6) and minimise the cost functional (4.6.5) by varying the parameters of the set

$$\wp_{c,\theta} = \{A_0, \ Q, \ m_0, \ m_1, \ k_0, \ k_1, \ c_0, \ c_1, \ b_0, \ b_1\} \,, \tag{4.6.7}$$

such that the resulting values of the parameter functions in (4.6.6) with (4.6.7) are in the set $\mathcal{P}_c$ for each $\theta$.

For further details and results, see Chapter 6 and Bökenheide et al. (2012b), Bökenheide et al. (2011), Bökenheide and Wolff (2012), Wolff et al. (2012c).

We summarise the optimisation scheme for creep in order to identify the material parameter set (4.6.2) required for model (4.4.1), (4.4.2) in Box 4.6.2.

### 4.6.2. Optimisation strategy for creep

- Given:
  - Experimental data sets

$$\left[(\epsilon_{L,exp})^i \ , \ (\epsilon_{D,exp})^i \ , \ (\epsilon_{c,exp})^i \ , \ (S_{exp})^i \ , \ (\theta_{exp})^i\right]_j \tag{4.6.8}$$

   where $j = 1, \ldots, N_{exp}$ and $i = 1, \ldots, N_j$
  - Define the set $\mathcal{P}_c$ with all possible combinations of parameters
  - Choose initial parameter set $\wp_{c0} \in \mathcal{P}_c$

- Optimisation:
  - Find optimal parameter set $\hat{\wp}_c \in \mathcal{P}_c$ fulfilling

$$\Phi(\epsilon^j_{c,exp}, S^j_{exp}, \epsilon^j_{c,cal}, S^j_{cal}, \hat{\wp}_c) = \min_{\wp \in \mathcal{P}_c} \sum_{j=1}^{N_{exp}} \Phi(\epsilon^j_{c,exp}, S^j_{exp}, \epsilon^j_{c,cal}, S^j_{cal}, \wp) , \tag{4.6.9}$$

   where $\epsilon^j_{c,cal} = (\epsilon^j_{c,cal\_strain}, \epsilon^j_{c,cal\_stress})$, by finding

$$\min_{\wp \in \mathcal{P}_c} \sum_{j=1}^{N_{exp}} \left[ \left( \sum_{i=1}^{N_j} \left| \epsilon^{i,j}_{c,exp} - \epsilon^{i,j}_{c,cal\_strain} \right|^2 \right)^{1/2} + \left( \sum_{i=1}^{N_j} \left| \epsilon^{i,j}_{c,exp} - \epsilon^{i,j}_{c,cal\_stress} \right|^2 \right)^{1/2} \right] \tag{4.6.10}$$

   by either
     * minimising the error considering one experimental test $j \in N_{exp}$ individually

   or

     * minimising the sum over all tests of the error between measured and calculated values

- Find $\wp_{c,\theta} = \{A_0, Q, m_0, m_1, k_0, k_1, c_0, c_1, b_0, b_1\}$ of the temperature-depending parameter functions in (4.6.6).

The optimal values of the parameter set $\hat{\wp}_c \in \mathcal{P}_c$ can be found using a a minimisation routine (e.g. via the MATLAB® routine fminsearch, fmincon), see Chapter 6 for details.

In Chapter 6 we will describe the performed experiments and the parameter identification in detail. The results of the experiments, the parameter identification together with the simulated creep strain will be presented.

For general aspects and further discussion on parameter identification we refer to Mahnken and Stein (1996),Mahnken (2004). A three-dimensional optimization procedure is presented in Yun and Shang (2011).

**Remark 4.6.1.** *(i) The cost functional in the algorithm 4.6.2 contains the strain- and stress-driven approach, see (4.6.10). Besides the difference between cal- culated and experimental strain, it is also possible to consider the error be- tween the stress $S_{exp}$ and $S_{cal}$. In this case, it possible to take either one, two or three errors in the cost functional into account.*

*(ii) When using the stress-driven approach, we consider the error between the measured and calculated creep strains in the cost functional. In addition to this, one can take the error of the difference of the measured longitudinal and transversal strains $\left(\epsilon_{L,exp}^{i,j} - \epsilon_{D,exp}^{i,j}\right)$ and the differences of the calculated strains using (4.4.18) into account.*

*(iii) In order to take phase transformations into account, we consider experiments using the different phases of the material. One first identifies the material parameters for each phase. After that, the total creep strain is modelled via a mixture rule using the single creep strains of each phase. See Chapter 6 for details.*

### 4.6.4 Parameter identification for TRIP

In the same manner as described in the case of creep in Section 4.6.3, an opti- misation routine for determining the TRIP parameter $\kappa$ (sometimes temperature- dependent) as well as the saturation function $\phi$ can be developed. In the case of the simultaneous occurrence of creep and TRIP, the creep strain $\epsilon_c$ is calculated in advance and is regarded as given.

Results for the parameter $\kappa$ of the performed parameter identification will be presented in Chapter 6.

# 5 Discretisation of the 3D model

In this chapter we focus on the discretisation and the numerical solution of the model presented in Chapter 2. The considered model is a coupled non-linear initial-boundary value problem of ordinary and partial differential equations. In order to be able to implement the model equations and thereby simulate the material behaviour of steel during heating and austenitisation, we have to handle the model numerically.

First of all, we consider the weak form of the partial differential equations in Section 5.1. The overall numerical solution scheme is presented in Section 5.2. In Section 5.3, we will deal with the application of the Finite Element Method.

In the case of inelastic material behaviour, the inelastic strain (i.e. here the creep strain) and the stress are coupled. We will develop an algorithm which determines (at each specific discrete time) the current creep strain and updates the stress tensor, see Section 5.4. The presented algorithm is a predictor-corrector approach. It was developed following the approach in Simo and Hughes (1998).

Results of the 3D simulations with the implemented model equations will be presented in Chapter 6.

Considering the first step of the solution scheme presented in Section 5.2.3, we follow Suhr (2010). The author handles the discretisation and implementation of the three-dimensional model equations for the case of classical plasticity and TRIP as well as phase transformations.

The modelling of material behaviour including TRIP and phase transformations as well as the implementation is handled in Wolff et al. (2000). In Schmidt et al. (2003), the authors present numerical investigations of the model.

## 5.1 Weak formulation

In order to simulate our model numerically, first we transform it into its weak form. In Section 5.3, we will apply the Finite Element Method to the weak form of the PDEs.

Obtaining the weak formulation of the heat equation and the deformation equation stated in Box 2.4.1, is a standard procedure. First, each equation is multiplied by a test function and then integrated over the domain $\Omega$, followed by some basic manipulations.

For further information we refer to Boettcher (2012), Boettcher et al. (2015), Suhr (2010) and Schmidt and Siebert (2005). Information about the used function spaces can be found e.g. in Zeidler (1990), Dautray and Lions (1992).

In the weak form of the equations, there arise some material parameters which depend on the solution itself. In the following, we assume that the material parameters evaluated at $\theta, p$ are in $L^\infty([0, T] \times \Omega)$.

**Remark 5.1.1** (Existence and uniqueness). *For discussion about existence and uniqueness of weak solutions to a similar coupled problem as the one presented here, we refer to Suhr (2010) and the references therein.*

*A thermo-elasto plastic problem with phase transitions in TRIP steels is handled in detail in Boettcher (2012) and Boettcher et al. (2015). In Boettcher (2012), the author proves existence and uniqueness of a solution of the corresponding initial-boundary value problem.*

*In Han and Reddy (1999), the authors deal with classical plasticity in the variational formulation. They also handle the numerical analysis of the variational problems.*

## 5.1.1 Weak form of heat equation

We consider the heat equation in (2.1.45) for $t \in (0, T_1)$ with Dirichlet boundary conditions

$$\theta = \vartheta \quad \text{on } \partial\Omega \times (0, T_1) . \tag{5.1.1}$$

We define the following function spaces:

$$V := H^{1,2}(\Omega) , \qquad \mathring{V} := \left\{ v \mid v \in H^{1,2}(\Omega) : v = 0 \text{ on } \partial\Omega \right\}, \tag{5.1.2a}$$

$$\mathcal{V} := L^2(0, T; V) , \qquad \mathring{\mathcal{V}} := L^2(0, T; \mathring{V}) , \tag{5.1.2b}$$

as well as

$$\mathcal{W} := \left\{ f \mid f \in \mathcal{V}, \frac{df}{dt} \in \mathcal{V}^* \right\}, \qquad \mathring{\mathcal{W}} := \left\{ f \mid f \in \mathring{\mathcal{V}}, \frac{df}{dt} \in \mathring{\mathcal{V}}^* \right\} . \tag{5.1.2c}$$

We assume that for $t \in (0, T_1)$ the Dirichlet boundary values $\vartheta$ have an extension to some function $\vartheta \in \mathcal{W}$ and $\varrho_0 > 0$, $p_i \in H^{1,2}(0, T; L^2(\Omega))$, $0 \le p_i \le 1$ *a.e.*, $c_d(\theta, p)$, $k_\theta(\theta, p)$, $L_i(\theta) \in L^\infty([0, T] \times \Omega)$.

We seek a solution $\theta \in \mathcal{W}$ with

$$\theta(x, 0) = \theta_0(x) \quad \text{a.e.,} \quad \theta_0 \in L^2(\Omega) , \tag{5.1.3}$$

such that $\theta \in \vartheta + \mathring{\mathcal{W}}$ for $t \in (0, T_1)$ and

$$\left\langle \varrho_0 c_d \frac{d\theta}{dt} \,\Big|\, \varphi \right\rangle + \int_\Omega (k_\theta \nabla\theta) \cdot \nabla\varphi \, dx = \int_\Omega \varrho_0 \sum_{i=2}^{M_p} L_i \dot{p}_i \varphi \, dx \tag{5.1.4}$$

for almost all $t \in (0, T_1)$ and for all $\varphi \in \mathring{V}$ where $\langle \cdot | \cdot \rangle$ denotes the dual pair in $V^* \times \mathring{V}$.

**Robin boundary conditions**

For $t \geq T_1$, we consider (2.1.45) with boundary conditions

$$- k_\theta(\theta, p)\nabla\theta \cdot n = \delta(\theta(x,t) - \theta_{ext_R}(t)) \quad \text{on} \quad \partial\Omega \times (T_1, T) \,. \tag{5.1.5}$$

We assume $\delta \in L^\infty(0,T; L^\infty(\partial\Omega))$ and $\theta_{ext_R} \in L^2(0,T; L^2(\partial\Omega))$. The equation has the following weak formulation: We search for a solution $\theta \in \mathcal{W}$ with

$$\theta(x,0) = \theta_0(x) \quad \text{a.e.,} \quad \theta_0 \in L^2(\Omega) \,, \tag{5.1.6}$$

so that

$$\left\langle \varrho_0 c_d \frac{d\theta}{dt} \,\middle|\, \varphi \right\rangle + \int_\Omega (k_\theta \nabla\theta) \cdot \nabla\varphi \, dx + \int_{\partial\Omega} \delta\theta\varphi \, ds$$

$$= \int_\Omega \varrho_0 \sum_{i=2}^{M_p} L_i \dot{p}_i \varphi \, dx + \int_{\partial\Omega} \delta\theta_{ext_R}\varphi \, ds \tag{5.1.7}$$

for almost all $t \in (T_1, T)$ and for all $\varphi \in V$.

## 5.1.2 Weak form of deformation equation

Now, we focus on the weak formulation of the deformation equation (2.1.46). Note that the inertia term may be neglected as we are usually not discussing situations with abruptly appearing or vanishing stresses (for a discussion of the inertia term, cf. Suhr (2010)). In this case, the partial differential equation under consideration changes its type from hyperbolic to elliptic. We denote the body force on the right-hand side of the equation as $\boldsymbol{f}_g$.

The weak formulation of the deformation equation with Neumann and (homogeneous) Dirichlet boundary conditions (cf. (2.1.46)) reads as follows.

We set

$$V_u := \left\{ \boldsymbol{w} \,\middle|\, \boldsymbol{w} \in [H^{1,2}(\Omega)]^3 : \boldsymbol{w} = 0 \text{ on } \Gamma_D \right\} \tag{5.1.8a}$$

$$\mathcal{V}_u := L^2(0,T; V_u) \tag{5.1.8b}$$

$$\mathcal{W}_u := \left\{ f \,\middle|\, f \in \mathcal{V}_u, \frac{df}{dt} \in L^2\big(0,T; [L^2(\Omega)]^3\big), \frac{d^2 f}{dt^2} \in \mathcal{V}_u^* \right\} \tag{5.1.8c}$$

and search for $\boldsymbol{u} \in \mathcal{W}_u$ with

$$\boldsymbol{u}(x,0) = 0 \quad \text{a.e.} \,, \quad \text{and} \quad \frac{d\boldsymbol{u}}{dt}(x,0) = 0 \quad \text{a.e.}$$

such that

$$\left\langle \varrho_0 \frac{d^2\boldsymbol{u}}{dt^2} \,\middle|\, \boldsymbol{v} \right\rangle + \int_\Omega 2\mu\boldsymbol{\varepsilon}(\boldsymbol{u}) : \boldsymbol{\varepsilon}(\boldsymbol{v}) \, dx + \int_\Omega \lambda \operatorname{div}(\boldsymbol{u}) \operatorname{div}(\boldsymbol{v}) \, dx =$$

$$\int_\Omega K\left(\frac{\varrho_0 - \varrho}{\varrho}\right) \boldsymbol{I} : \nabla\boldsymbol{v} \, dx + \int_\Omega \varrho_0 \boldsymbol{f}_g \cdot \boldsymbol{v} \, dx + \int_\Omega 2\mu\boldsymbol{\varepsilon}_{in} : \nabla\boldsymbol{v} \, dx + \int_{\Gamma_N} g_N \cdot \boldsymbol{v} \, ds$$

$$\tag{5.1.9}$$

for almost all $t \in (0,T)$ and for all $\boldsymbol{v} \in V_u$ where $\varrho_0 > 0$,
$\mu(\theta,p)$, $\lambda(\theta,p)$, $K(\theta,p)$, $\varrho(\theta,p) \in L^\infty([0,T] \times \Omega)$, $\varrho(\theta,p) \geq c > 0$ $a.e.$,
$\boldsymbol{f}_g \in L^2(0,T;[L^2(\Omega)]^3)$, $g_N \in L^2(0,T;[L^2(\Gamma_N)]^3)$, $\boldsymbol{\varepsilon}_{in} \in L^2(0,T;[L^2(\Omega)]^5)$ and $\langle \cdot | \cdot \rangle$
denotes the dual pair in $V_u^* \times V_u$.

## 5.2 Numerical solution scheme

Now, we will deal with the discretisation of the model presented in Box 2.4.1.
First, we discretise the solution interval $[0,T]$ into time steps. After that, we will
describe how we proceed in each time step in Section 5.2.3.

### 5.2.1 Time discretisation

First, we divide the considered time interval $[0,T]$ into discrete time steps:

$$t_0 := 0 < t_1 < \cdots < t_n < \cdots < t_{N_t} := T \;. \tag{5.2.1}$$

We consider the model presented in Section 2.4. At a specific discrete time $t_n$,
where $n \in \{1, \ldots, N_t\}$ our aim is to determine

$$\theta^n, \quad p^n, \quad \boldsymbol{u}^n, \quad \boldsymbol{\varepsilon}^n, \quad \boldsymbol{\sigma}^n, \quad \boldsymbol{\varepsilon}_{in}^n, \quad \boldsymbol{X}^n,$$

which each denote functions on $\Omega$ (for simplicity, we suppress the dependence
on $x$ in the notation). In the following, the superscript $n$ always denotes the
approximated discrete value of the variable at time $t_n$, thus for instance $\theta^n \approx \theta(t_n)$.

We define the $n$-th time step size as $\tau_n := t_n - t_{n-1}$. The time derivative is
approximated by the difference quotient

$$\dot{\theta}(t_n) \approx \dot{\theta}^n = \frac{\theta^n - \theta^{n-1}}{\tau_n} \;. \tag{5.2.2}$$

Furthermore, we describe the so-called $\vartheta$-method used for the time discretisation.
We consider the heat equation, summarize the equation as

$$\frac{\partial \theta}{\partial t} = F\left(x, t, \theta, \frac{\partial \theta}{\partial x}, \frac{\partial^2 \theta}{\partial x^2}\right), \tag{5.2.3}$$

and discretise it by

$$\frac{\theta^n - \theta^{n-1}}{\tau_n} = \vartheta F\left(x, t_n, \theta^n, \frac{\partial \theta^n}{\partial x}, \frac{\partial^2 \theta^n}{\partial x^2}\right) + (1-\vartheta) F\left(x, t_{n-1}, \theta^{n-1}, \frac{\partial \theta^{n-1}}{\partial x}, \frac{\partial^2 \theta^{n-1}}{\partial x^2}\right)$$
$$+ G\left(x, t_{n-1}, \theta^{n-1}\right) \tag{5.2.4}$$

Thus, choosing $\vartheta = 0$, (5.2.4) corresponds to an explicit Euler scheme, choosing
$\vartheta = 1$ to an implicit Euler scheme. By setting $\vartheta = \frac{1}{2}$, we obtain a Crank-Nicolson
scheme.

## 5.2.2 Spatial discretisation

We choose a shape regular, simplicial triangulation $\mathcal{T}_h$ for the discretisation of the region $\Omega$. Considering the 3D case, our triangulation consists of tetrahedrons of maximal edge length $h$. We use piecewise polynomial Lagrange Finite Elements $\varphi$ for the definition of the corresponding Finite Element space. We set:

$$V_h = \text{span}\{\varphi_1, \ldots, \varphi_N\} \subset V \tag{5.2.5}$$

Next, we define a Finite Element space of dimension $3N$ for the three-dimensional vector field $\boldsymbol{u}^n$:

$$V_h^3 = \text{span}\left\{ \begin{pmatrix} \varphi_i \\ 0 \\ 0 \end{pmatrix}, \begin{pmatrix} 0 \\ \varphi_i \\ 0 \end{pmatrix}, \begin{pmatrix} 0 \\ 0 \\ \varphi_i \end{pmatrix} \right\}_{i=1,\ldots,N}$$
$$= \text{span}\{\boldsymbol{\phi}_1, \ldots, \boldsymbol{\phi}_{3N}\} \subset V_u \, . \tag{5.2.6}$$

In order to define a Finite Element space for the tensorial quantities, we exploit the fact that in our setting all tensors are symmetric. This saves a considerable amount of memory.

We set

$$V_h^6 = \text{span}\{\Phi_{jk}^i \, , \; j, k, = 1, 2, 3, \; j \leq k\}_{i=1,\ldots,N} \, , \tag{5.2.7}$$

where the elements are composed as follows:

$$\Phi_{11}^i = \begin{pmatrix} \varphi_i & 0 & 0 \\ 0 & 0 & 0 \\ 0 & 0 & 0 \end{pmatrix} \quad , \quad \Phi_{12}^i = \frac{1}{\sqrt{2}} \begin{pmatrix} 0 & \varphi_i & 0 \\ \varphi_i & 0 & 0 \\ 0 & 0 & 0 \end{pmatrix} \, , \tag{5.2.8}$$

and so forth. Here, the factor for the non-diagonal elements arises in order to ensure that $\left(\Phi_{jk}^i, \Phi_{jk}^i\right)_{L^2(\Omega)} = \int_\Omega \varphi_i \varphi_j \, dx$ holds.

In the case of tensors which are symmetric as well as traceless, it is sufficient to use a Finite Element space of dimension $5N$,

$$V_h^5 = \text{span}\{\Phi_{jk}^i \, , \; j = 1, 2, \; k, = 1, 2, 3, \; j \leq k\}_{i=1,\ldots,N} \, . \tag{5.2.9}$$

## 5.2.3 Overall solution scheme

Our underlying model is a coupled model of partial and ordinary differential equation. In order to handle the PDEs, i.e. the heat equation and the deformation equation, we apply the Finite Element Method. The steps are summarized in Box 5.2.1.

### 5.2.1. Numerical solution scheme

Start with initial values at $t_0$: $\theta^0, p^0, \boldsymbol{u}^0, \boldsymbol{\sigma}^0, \boldsymbol{\varepsilon}_{in}^0, \boldsymbol{X}^0$.
Compute the current quantities for $t_n$, all quantities for $t_{n-1}$ are known.

**I) Calculate $\theta^n, p^n, \boldsymbol{u}^n$:**

- obtain $\theta^n$ by solving the heat equation (5.1.7) via the Finite Element Method using the material parameters evaluated at $\theta^{n-1}, p^{n-1}$

- solve ODEs for phase fractions $p^n$ via Euler scheme

- compute $\boldsymbol{u}^n$ by solving the deformation equation via Finite Element Method, using former values of $\boldsymbol{\varepsilon}_{in}^{n-1}$

- obtain the strain tensor $\boldsymbol{\varepsilon}(\boldsymbol{u}^n)$

**II) Calculate inelastic quantities $\boldsymbol{\varepsilon}_{in}^n, \boldsymbol{X}^n, \boldsymbol{\sigma}^n$:**

- compute the inelastic strain $\boldsymbol{\varepsilon}_{in}^n$ and the back stress $\boldsymbol{X}^n$ using a trial stress $\boldsymbol{\sigma}_t^n$ by means of the algorithm presented in Section 5.4

- update the stress tensor $\boldsymbol{\sigma}^n$

The details of the different steps in the solution scheme will be explained in the following. Section 5.3 will handle the first part of the solution scheme, in Section 5.4, we will present the algorithm that will be used to calculate the inelastic quantities.

## 5.3 Application of the Finite Element Method

Now, we apply the Finite Element Method to our model equations. Details and further information about the method can be found in Braess (1992), Knabner and Angermann (2000), e.g.

### 5.3.1 Discretised heat equation

First, we focus on solving the heat conduction equation presented in Section 2.1.4. Our aim is to calculate the approximate solution $\theta^n \in V_h$ to the exact solution $\theta(t_n) \in V$. We set

$$\theta^n = \sum_{i=1}^{N} \theta_i^n \varphi_i \,, \ \in V_h \,, \tag{5.3.1}$$

thus we have to calculate the coefficient vector $(\theta_1^n, \ldots, \theta_N^n)$.

Next, we apply the Crank-Nicolson scheme to the weak form of the heat equation (5.1.7), choosing $\varphi_j$ with $j = 1, \ldots, N$ as test functions (cf. (5.2.4)). We have

$$\int_\Omega \varrho_0 c_{d,n-1} \frac{\theta^n - \theta^{n-1}}{\tau_n} \varphi_j \, dx + \vartheta \int_\Omega k_{\theta,n-1} \nabla\theta^n \cdot \nabla\varphi_j \, dx + \int_{\Gamma_R} \delta\theta^n \varphi_j \, ds =$$

$$(1 - \vartheta) \int_\Omega k_{\theta,n-1} \nabla\theta^{n-1} \cdot \nabla\varphi_j \, dx + \int_\Omega \varrho_0 \sum_{k=2}^{M_p} L_k(\theta^{n-1}) \dot{p}^{\,n-1} \varphi_j \, dx + \int_{\Gamma_R} \delta\theta_{ext}^{n-1} \varphi_j \, ds$$

$$\text{for } j = 1, \ldots, N , \qquad (5.3.2)$$

where $c_{d,n-1} = c_d(\theta^{n-1}, p^{n-1})$, $k_{\theta,n-1} = k_\theta(\theta^{n-1}, p^{n-1})$. We apply the Galerkin method for the spatial discretisation of the time discretised heat equation. Inserting (5.3.1) into (5.3.2), we obtain a system of linear equations of dimension $N \times N$:

$$\sum_{i=1}^N \left( \frac{\varrho_0}{\tau_n} \int_\Omega c_{d,n-1} \varphi_i \varphi_j \, dx + \vartheta \int_\Omega k_{\theta,n-1} \nabla\varphi_i \cdot \nabla\varphi_j \, dx + \int_{\Gamma_R} \delta\varphi_i \varphi_j \, ds \right) \theta_i^n =$$

$$\sum_{i=1}^N \left( \frac{\varrho_0}{\tau_n} \int_\Omega c_{d,n-1} \varphi_i \varphi_j \, dx + (1 - \vartheta) \int_\Omega k_{\theta,n-1} \nabla\varphi_i \cdot \nabla\varphi_j \, dx \right) \theta_i^{n-1} +$$

$$+ \int_\Omega \varrho_0 \sum_{k=2}^{M_p} L_k(\theta^{n-1}) \dot{p}_k^{n-1} \varphi_j \, dx + \int_{\Gamma_R} \delta\theta_{ext}^{n-1} \varphi_j \, ds$$

$$\text{for } j = 1, \ldots, N , \qquad (5.3.3)$$

which represents the fully discretised form of the heat equation. Solving (5.3.3), we obtain the coefficients $(\theta_1^n, \ldots, \theta_N^n)$ which yields $\theta^n$, i.e. the approximate solution to $\theta(t_n)$ in $V_h$.

## 5.3.2 Phase transitions

The evolution of the phase fractions is described by ordinary differential equations, see Section 2.1.3 and 2.3. The corresponding ODEs are solved by means of an implicit Euler scheme. The calculations are conducted in each degree of freedom separately.

Let $M_p$ denote the number of phases. We use the following ansatz of ordinary differential equations for the phase fractions $p_{k,i}$:

$$\dot{p}_{k,i} = F_{p_{k,i}}(\theta, p_i) , \quad \text{for } k = 1, \ldots, M_p , \ i = 1, \ldots, N , \qquad (5.3.4)$$

where $p_i = (p_{1,i}, \ldots, p_{M_p,i})$. The current values of the phase fractions $p_{k,i}^n$ in the single nodes follow by

$$p_{k,i}^n = p_{k,i}^{n-1} + \tau_n F_{p_{k,i}}^n(\theta^n, p_i^{n-1}) , \qquad (5.3.5)$$

for $k = 1, \ldots, M_p$ and $p_i^n = (p_{1,i}^n, \ldots, p_{M_p,i}^n)$.

### 5.3.3 Discretised deformation equation

Next, our aim is to determine the displacement $\boldsymbol{u}^n$ by solving the deformation equations presented in Section 2.1.5. We consider the weak form of the deformation equation (5.1.9). In the following, we neglect the inertia term and handle the so-called quasi-static case. Hence, the partial differential equation under consideration changed its type from hyperbolic to elliptic:

$$\int_\Omega 2\mu_n \boldsymbol{\varepsilon}(\boldsymbol{u}^n) : \boldsymbol{\varepsilon}(\boldsymbol{\phi}_j)\, dx \;+\; \int_\Omega \lambda_n \operatorname{div}(\boldsymbol{u}^n)\operatorname{div}(\boldsymbol{\phi}_j)\, dx =$$

$$\int_\Omega \varrho_0 \boldsymbol{f}_g^n \cdot \boldsymbol{\phi}_j\, dx + \int_\Omega K_n \left( \frac{\varrho_0 - \varrho_n}{\varrho_n} \right) \boldsymbol{I} : \nabla \boldsymbol{\phi}_j\, dx + \int_\Omega 2\mu_n \boldsymbol{\varepsilon}_{in}^{n-1} : \nabla \boldsymbol{\phi}_j\, dx$$

$$\text{for } j = 1,\dots,N\,, \qquad (5.3.6)$$

where the lower subscript $n$ denotes the value of a material parameter in $(\theta^n, p^n)$. As the equation is stationary, only the spatial discretisation is required. We proceed in the same way as for the heat equation: We set

$$\boldsymbol{u}^n = \sum_{i=1}^{3N} u_i^n \boldsymbol{\phi}_i \; \in V_h^3\,, \qquad (5.3.7)$$

insert (5.3.7) into (5.3.6) and obtain

$$\sum_{i=1}^{3N} \left( \int_\Omega 2\mu_n \boldsymbol{\varepsilon}(\boldsymbol{\phi}_i) : \boldsymbol{\varepsilon}(\boldsymbol{\phi}_j)\, dx + \int_\Omega \lambda_n \operatorname{div}(\boldsymbol{\phi}_i)\operatorname{div}(\boldsymbol{\phi}_j)\, dx \right) u_i^n =$$

$$\int_\Omega \varrho_0 \boldsymbol{f}_g^n \cdot \boldsymbol{\phi}_j\, dx + \int_\Omega K_n \left( \frac{\varrho_0 - \varrho_n}{\varrho_n} \right) \boldsymbol{I} : \nabla \boldsymbol{\phi}_j\, dx + \int_\Omega 2\mu_n \boldsymbol{\varepsilon}_{in}^{n-1} : \nabla \boldsymbol{\phi}_j\, dx$$

$$\text{for } j = 1,\dots,3N\,. \quad (5.3.8)$$

Altogether, we get a linear system of equations of dimension $3N \times 3N$. Its solution, i.e. the coefficient vector $(u_1^n,\dots,u_{3N}^n)$ yields the sought approximation $\boldsymbol{u}^n$ of $\boldsymbol{u}(t_n)$ in the FE space $V_h^3$.

### 5.3.4 Strain and stress tensor

Having determined $\boldsymbol{u}^n$, the total strain tensor $\boldsymbol{\varepsilon}(\boldsymbol{u}^n)$ follows by

$$\boldsymbol{\varepsilon}(\boldsymbol{u}^n) = \frac{1}{2}\left( \nabla \boldsymbol{u}^n + (\nabla \boldsymbol{u}^n)^T \right)\,. \qquad (5.3.9)$$

Note that, as the strain tensor contains the gradient of the displacement, its components $\epsilon_{kl}$ ($k,l = 1,2,3$) are not element of the Finite Element space. Therefore, we project the strain tensor component by component onto $V_h$:

$$\tilde{\epsilon}_{kl} := P_{V_h}(\epsilon_{kl})\,, \quad k,l = 1,2,3,\ k \le l\,. \qquad (5.3.10)$$

As we consider the projection onto $V_h$, it holds that

$$\tilde{\epsilon}_{kl} \in V_h \ , \ \tilde{\epsilon}_{kl} = \sum_{i=1}^{N} (\tilde{\epsilon}_{kl})_i \, \varphi_i \quad \text{and} \quad \int_{\Omega} \tilde{\epsilon}_{kl} \varphi_i \, dx = \int_{\Omega} \epsilon_{kl} \varphi_i \, dx \ . \tag{5.3.11}$$

Inserting the first equation of (5.3.11) into the second, yields a linear system of equations,

$$\sum_{i=1}^{N} \left( \int_{\Omega} \varphi_i \varphi_j \, dx \right) (\tilde{\epsilon}_{kl})_i = \int_{\Omega} \epsilon_{kl} \varphi_j \, dx \ , \quad j = 1, \ldots, N \ , \tag{5.3.12}$$

which can be used to determine the coefficients $(\tilde{\epsilon}_{kl})_i$. For simplicity, we suppress the tilde in the notation. Whenever the strain tensor is mentioned, its projection is meant.

The calculated projection of the strain tensor onto the Finite Element space is used to compute the stress tensor $\boldsymbol{\sigma}^n$ via

$$\boldsymbol{\sigma}^n = 2\mu_n \left( \boldsymbol{\varepsilon}(\boldsymbol{u}^n) - \boldsymbol{\varepsilon}_{in}^{n-1} \right) + \left( \lambda_n \text{tr}(\boldsymbol{\varepsilon}(\boldsymbol{u}^n)) - K_n \left( \frac{\varrho_0 - \varrho_n}{\varrho_n} \right) \right) \boldsymbol{I} \ . \tag{5.3.13}$$

This concludes part one of the algorithm in Box 5.2.1. In the next step, our aim is to determine the current inelastic strain $\boldsymbol{\varepsilon}_{in}^n$ as well as the updated value of the stress tensor $\boldsymbol{\sigma}^n$ by inserting $\boldsymbol{\varepsilon}_{in}^n$ in (5.3.13). Section 5.4 will handle the numerical algorithm in detail.

**Remark 5.3.1** (Numerical quadrature). *In order to compute integrals as $\int_{\Omega} f(x) \varphi_i \, dx$, quadrature formulas are used to calculate the integrals approximately. In Finite Element Methods, numerical integration is realised by looping over all grid elements and using a quadrature formula on each element (cf. Schmidt and Siebert (2005)). Here, we use Gaussian quadrature. For further information, we refer to Schmidt and Siebert (2005) and Ciarlet (2002), e.g.*

## 5.4 Numerical algorithm for the calculation of inelastic quantities

In this section, we will deal in some detail with the second step of the overall numerical scheme presented in Section 5.2.3 in order to calculate the inelastic quantities, see Box 5.2.1. The sought quantities are space-dependent. We conduct the calculations in each degree of freedom separately. Therefore, this step can be understood as a post-processing after solving the deformation equation. The presented algorithm is a predictor-corrector approach (cf. Simo and Hughes (1998)).

In the case of inelastic material behaviour, we have to calculate the inelastic strain, back stress as well as an updated value of the stress. The implemented algorithm which we present in the following was developed based on the algorithms in Simo and Hughes (1998), Suhr (2010) and Wolff et al. (2011b).

We will deal with our model problem considering creep material behaviour. Hence, the inelastic strain corresponds to the creep strain, i.e. $\varepsilon_{in} = \varepsilon_c$. We will reformulate the material laws by introducing a scalar value which we will refer to as a *creep multiplier* following the terminology in the case of classical plastic material behaviour, see Section 5.4.1. We are thereby able to reduce our 3D problem to a 1D problem depending only on the scalar multiplier.

We proceed as follows: First, we calculate a so-called *trial stress* tensor. After that, we are able to formulate an equation for the current value of the creep strain tensor using the trial stress. We will solve for the creep multiplier first and then update the remaining quantities. After that, we correct the stress tensor by means of the updated value of the creep strain.

## 5.4.1 Algorithm for creep

Our aim is to solve the evolution equation for the creep strain introduced in Chapter 2, see Box 2.4.1. In the following, we set $D_c = 1$. We use the model by Armstrong-Frederick (see (2.2.30), (2.2.34)):

$$\dot{\varepsilon}_c = \frac{3}{2} A(\theta) \left( \sqrt{\frac{3}{2}} \| \boldsymbol{\sigma}^* - \boldsymbol{X}_c^* \| \right)^{m(\theta)-1} (\boldsymbol{\sigma}^* - \boldsymbol{X}_c^*) \, s_c^{k(\theta)} \tag{5.4.1}$$

$$\dot{\boldsymbol{X}}_c = \frac{2}{3} c_c(\theta) \, \dot{\varepsilon}_c - b_c(\theta, \boldsymbol{\sigma}) \, \boldsymbol{X}_c \, \dot{s}_c + \frac{\mathrm{d} c_c}{\mathrm{d}\theta} \frac{\dot{\theta}}{c_c(\theta)} \, \boldsymbol{X}_c \,, \tag{5.4.2}$$

with $A > 0$, $m > 0$, $c_c > 0$, $b_c \geq 0$.

First we write equation (5.4.1) in the following form:

$$\dot{\varepsilon}_c = \sqrt{\frac{3}{2}} A(\theta) \left( \sqrt{\frac{3}{2}} \| \boldsymbol{\sigma}^* - \boldsymbol{X}_c^* \| \right)^{m(\theta)} \frac{\boldsymbol{\sigma}^* - \boldsymbol{X}_c^*}{\| \boldsymbol{\sigma}^* - \boldsymbol{X}_c^* \|} \, s_c^{k(\theta)} \,. \tag{5.4.3}$$

Next, we reformulate the evolution equation by introducing the scalar $\gamma_c$ which we will refer to as a creep multiplier. In contrast to the case of classical plasticity, in our case there is no yield function given that characterises the multiplier. Here, the multiplier can directly be stated.

We define the creep multiplier as

$$\gamma_c := \sqrt{\frac{3}{2}} A(\theta) \left( \sqrt{\frac{3}{2}} \| \boldsymbol{\sigma}^* - \boldsymbol{X}_c^* \| \right)^{m(\theta)} s_c^{k(\theta)} \,. \tag{5.4.4}$$

Now, we are able to write the evolution equation of the creep strain in the following form:

$$\dot{\varepsilon}_c = \gamma_c \frac{\boldsymbol{\sigma}^* - \boldsymbol{X}_c^*}{\| \boldsymbol{\sigma}^* - \boldsymbol{X}_c^* \|} \,, \tag{5.4.5}$$

(following the notation for classical plasticity, cf. Section 2.2.1). For the evolution of the accumulated creep strain it follows that

$$\dot{s}_c = \sqrt{\frac{2}{3}}\,\|\dot{\varepsilon}_c\| = \sqrt{\frac{2}{3}}\,\gamma_c \ . \tag{5.4.6}$$

In the following, we will describe the discretisation of the model equations as well as the numerical algorithm. Our aim is to compute the current values of the creep strain tensor $\varepsilon_c^n$, the back stress $\boldsymbol{X}_c^n$ as well as of the stress tensor $\boldsymbol{\sigma}^n$ for $t_n$. The values of $\theta^n, p^n$ and $\boldsymbol{u}^n$ as well as all values from former time steps are assumed to be known at this step of the algorithm (cf. Box 5.2.1). The total strain tensor $\varepsilon^n$ follows by (5.3.9) and can be considered as given, too.

The current value of the stress tensor $\boldsymbol{\sigma}^n$ required for the calculation of $\varepsilon_c^n$, is not yet known at this step of the algorithm. Instead of the current value, we will use a trial stress for the calculation of the inelastic strain. Having determined the creep strain, the stress tensor is updated. This is called a 'correction' of the stress tensor.

We summarize the steps of the algorithm in Box 5.4.1.

---

**5.4.1. Numerical algorithm for the calculation of inelastic quantities**

Start with initial values $\theta^0, p^0, \boldsymbol{u}^0, \boldsymbol{\sigma}^0, \varepsilon^0, \varepsilon_c^0, \boldsymbol{X}_c^0$. Compute the current quantities at time $t_n$. All quantities for $t_{n-1}$ are known.

The current temperature $\theta^n$, the phase fractions $p^n$ as well as the deformation $\boldsymbol{u}^n$ are given by step I) of the algorithm in Box 5.2.1.

**Step II) : Calculate inelastic quantities** $\varepsilon_c^n, \boldsymbol{X}_c^n, \boldsymbol{\sigma}^n$

1. Calculate (the deviator of) the trial stress $\boldsymbol{\sigma}_t^{*n}$

2. Formulate equation for creep multiplier $\gamma_c^n$ and solve it numerically

3. Update inelastic quantities, i.e. $\varepsilon_c^n, s_c^n, \boldsymbol{X}_c^n$

4. Correct stress tensor $\boldsymbol{\sigma}^n$ using the updated value of $\varepsilon_c^n$

---

In the next subsections, we focus on the specific cases of creep with a back stress as well as creep without back stress, see Section 5.4.2 and 5.4.3.

## 5.4.2 Algorithm for creep with a back stress

Our aim is to calculate the current value of the creep strain, of the back stress and of the stress tensor. Therefore, we will consider the reformulated evolution equation for the creep strain tensor (5.4.5) and solve the problem for the creep multiplier $\gamma_c^n$ first.

**Step 1: Trial stress**

First, we calculate a so-called 'trial stress' taking values of the former time step for the inelastic strain into account. The trial stress tensor $\boldsymbol{\sigma}_t^n$ is defined as

$$\boldsymbol{\sigma}_t^n := 2\mu(\theta^n, p^n)(\boldsymbol{\varepsilon}^n - \boldsymbol{\varepsilon}_c^{n-1}) + \lambda(\theta^n, p^n)(\mathrm{tr}(\boldsymbol{\varepsilon}^n))\boldsymbol{I} - K(\theta^n, p^n)\left(\frac{\varrho_0 - \varrho(\theta^n, p^n)}{\varrho(\theta^n, p^n)}\right)\boldsymbol{I}.$$
(5.4.7)

Thus, with the equation for the stress tensor (5.3.13) and

$$\boldsymbol{\varepsilon}^{*n} := (\boldsymbol{\varepsilon}^n)^* = \boldsymbol{\varepsilon}^n - \frac{1}{3}\mathrm{tr}(\boldsymbol{\varepsilon}^n)\boldsymbol{I},$$
(5.4.8)

we obtain

$$\begin{aligned}
\boldsymbol{\sigma}^n = {}& 2\mu(\theta^n, p^n)(\boldsymbol{\varepsilon}^n - \boldsymbol{\varepsilon}_c^n) + \lambda(\theta^n, p^n)(\mathrm{tr}(\boldsymbol{\varepsilon}^n))\boldsymbol{I} - K(\theta^n, p^n)\left(\frac{\varrho_0 - \varrho(\theta^n, p^n)}{\varrho(\theta^n, p^n)}\right)\boldsymbol{I} \\
= {}& 2\mu(\theta^n, p^n)(\boldsymbol{\varepsilon}^{*n} - \boldsymbol{\varepsilon}_c^{n-1}) + \lambda(\theta^n, p^n)(\mathrm{tr}(\boldsymbol{\varepsilon}^n))\boldsymbol{I} + 2\mu(\theta^n, p^n)(\boldsymbol{\varepsilon}_c^{n-1} - \boldsymbol{\varepsilon}_c^n) \\
& + \frac{2}{3}\mu(\theta^n, p^n)(\mathrm{tr}(\boldsymbol{\varepsilon}^n - \boldsymbol{\varepsilon}_c^n))\boldsymbol{I} - K(\theta^n, p^n)\left(\frac{\varrho_0 - \varrho(\theta^n, p^n)}{\varrho(\theta^n, p^n)}\right)\boldsymbol{I},
\end{aligned}$$
(5.4.9)

exploiting the fact that the inelastic strain is traceless. The current value of the time derivative of $\boldsymbol{\varepsilon}_c$ is approximated by the difference quotient

$$\dot{\boldsymbol{\varepsilon}}_c(t_n) \approx \dot{\boldsymbol{\varepsilon}}_c^n = \frac{\boldsymbol{\varepsilon}_c^n - \boldsymbol{\varepsilon}_c^{n-1}}{t_n - t_{n-1}}.$$
(5.4.10)

Using this approximation and (5.4.7), we finally obtain the *corrected stress* tensor as

$$\begin{aligned}
\boldsymbol{\sigma}^n = {}& \boldsymbol{\sigma}_t^{*n} - 2\mu(\theta^n, p^n)\tau_n(\dot{\boldsymbol{\varepsilon}}_c)^n + \left(\lambda(\theta^n, p^n) + \frac{2}{3}\mu(\theta^n, p^n)\right)\mathrm{tr}(\boldsymbol{\varepsilon}^n)\boldsymbol{I} \\
& - K(\theta^n, p^n)\left(\frac{\varrho_0 - \varrho(\theta^n, p^n)}{\varrho(\theta^n, p^n)}\right)\boldsymbol{I},
\end{aligned}$$
(5.4.11)

where $\tau_n := t_n - t_{n-1}$ and $\boldsymbol{\sigma}_t^{*n}$ states the deviatoric part of the current stress tensor, i.e $\boldsymbol{\sigma}_t^{*n} := (\boldsymbol{\sigma}^n)^*$. Hence, the deviatoric part of the corrected stress tensor is given by

$$\boldsymbol{\sigma}^{*n} = \boldsymbol{\sigma}_t^{*n} - 2\mu(\theta^n, p^n)\tau_n(\dot{\boldsymbol{\varepsilon}}_c)^n.$$
(5.4.12)

If we model the evolution of the creep strain with a back stress (see (5.4.1)–(5.4.2)), an additional evolution equation for the back stress $\boldsymbol{X}_c^n$ is required.

First, we introduce the definition of the so-called *effective stress* by

$$\boldsymbol{\xi}_c^n := \boldsymbol{\sigma}^{*n} - \boldsymbol{X}_c^{*n} \ . \tag{5.4.13}$$

Using (5.4.13) and (5.4.5), we have

$$(\dot{\boldsymbol{\varepsilon}}_c)^n = \gamma_c^n \frac{\boldsymbol{\xi}_c^n}{\|\boldsymbol{\xi}_c^n\|} \ . \tag{5.4.14}$$

Thus, the deviator of the corrected stress tensor introduced in (5.4.12) is given by

$$\boldsymbol{\sigma}^{*n} = \boldsymbol{\sigma}_t^{*n} - 2\mu(\theta^n, p^n)\tau_n\gamma_c^n \frac{\boldsymbol{\xi}_c^n}{\|\boldsymbol{\xi}_c^n\|} \ . \tag{5.4.15}$$

**Step 2: Creep strain**

In order to calculate the creep strain, we have to determine the other unknowns first. In the next steps, we will calculate the required values in the following order:

$$\gamma_c^n \ , \ \|\boldsymbol{\xi}_c^n\| \ , \ \boldsymbol{\xi}_c^n \ , \ \boldsymbol{X}_c^n \ , \ \boldsymbol{\sigma}^{*n} \ , \ \boldsymbol{\varepsilon}_c^n \ , \ s_c^n \ .$$

Taking

$$(\dot{s}_c)^n = \sqrt{\frac{2}{3}}\|(\dot{\boldsymbol{\varepsilon}}_c)^n\| = \sqrt{\frac{2}{3}}\gamma_c^n \ , \tag{5.4.16}$$

and (5.4.2) into account, the discretised equation for the back stress $\boldsymbol{X}_c^n$ is given by:

$$\boldsymbol{X}_c^n = \boldsymbol{X}_c^{n-1} + \frac{2}{3}\tau_n c_c(\theta^n)\,\gamma_c^n \frac{\boldsymbol{\xi}_c^n}{\|\boldsymbol{\xi}_c^n\|} - \tau_n b_c(\theta^n)\,\boldsymbol{X}_c^n \sqrt{\frac{2}{3}}\gamma_c^n + \Delta c_c(\theta^n)\boldsymbol{X}_c^n \tag{5.4.17}$$

where $\Delta c_c(\theta^n) = \frac{c_c(\theta^n)-c_c(\theta^{n-1})}{c_c(\theta^n)}$. We reformulate this equation with respect to $\boldsymbol{X}_c^n$:

$$\boldsymbol{X}_c^n \left(1 + \tau_n b_c(\theta^n) \sqrt{\frac{2}{3}}\gamma_c^n - \Delta c_c(\theta^n)\right) = \boldsymbol{X}_c^{n-1} + \frac{2}{3}\tau_n c_c(\theta^n)\,\gamma_c^n \frac{\boldsymbol{\xi}_c^n}{\|\boldsymbol{\xi}_c^n\|} , \tag{5.4.18}$$

which leads to

$$\boldsymbol{X}_c^n = \underbrace{\left(1 + \tau_n b_c(\theta^n) \sqrt{\frac{2}{3}}\gamma_c^n - \Delta c_c(\theta^n)\right)^{-1}}_{=:g(\gamma_c^n)} \left(\boldsymbol{X}_c^{n-1} + \frac{2}{3}\tau_n c_c(\theta^n)\,\gamma_c^n \frac{\boldsymbol{\xi}_c^n}{\|\boldsymbol{\xi}_c^n\|}\right) \ .$$

$$\tag{5.4.19}$$

Next, we formulate an equation for $\|\boldsymbol{\xi}_c^n\|$. We insert (5.4.15) and (5.4.19) into (5.4.13) which yields:

$$\boldsymbol{\xi}_c^n = \boldsymbol{\sigma}_t^{*n} - 2\mu(\theta^n, p^n)\tau_n\gamma_c^n \frac{\boldsymbol{\xi}_c^n}{\|\boldsymbol{\xi}_c^n\|} - g(\gamma_c^n) \left(\boldsymbol{X}_c^{n-1} + \frac{2}{3}\tau_n c_c(\theta^n)\,\gamma_c^n \frac{\boldsymbol{\xi}_c^n}{\|\boldsymbol{\xi}_c^n\|}\right) \tag{5.4.20}$$

$$= \boldsymbol{\sigma}_t^{*n} - g(\gamma_c^n)\boldsymbol{X}_c^{n-1} - \tau_n\gamma_c^n \frac{\boldsymbol{\xi}_c^n}{\|\boldsymbol{\xi}_c^n\|} \left(2\mu(\theta^n, p^n) + \frac{2}{3}c_c(\theta^n)g(\gamma_c^n)\right) \ . \tag{5.4.21}$$

We reformulate this equation:

$$\boldsymbol{\xi}_c^n \left(1 + \tau_n \gamma_c^n \frac{1}{\|\boldsymbol{\xi}_c^n\|} \left(2\mu(\theta^n, p^n) + \frac{2}{3} c_c(\theta^n) g(\gamma_c^n)\right)\right) = \boldsymbol{\sigma}_t^{*n} - g(\gamma_c^n) \boldsymbol{X}_c^{n-1} . \quad (5.4.22)$$

Applying the norm to equation (5.4.22), we obtain

$$\|\boldsymbol{\sigma}_t^{*n} - g(\gamma_c^n) \boldsymbol{X}_c^{n-1}\| = \|\boldsymbol{\xi}_c^n\| \left| \left(1 + \tau_n \gamma_c^n \frac{1}{\|\boldsymbol{\xi}_c^n\|} \left(2\mu(\theta^n, p^n) + \frac{2}{3} c_c(\theta^n) g(\gamma_c^n)\right)\right)\right|$$

$$= \left| \left(\|\boldsymbol{\xi}_c^n\| + \tau_n \gamma_c^n \left(2\mu(\theta^n, p^n) + \frac{2}{3} c_c(\theta^n) g(\gamma_c^n)\right)\right)\right| \quad (5.4.23)$$

Next, we formulate a second equation for the norm $\|\boldsymbol{\xi}_c^n\|$. We use the equation for the creep multiplier (see (5.4.4) and (5.4.16)),

$$\gamma_c^n = \sqrt{\frac{3}{2}} A(\theta^n) \left(\sqrt{\frac{3}{2}} \|\boldsymbol{\xi}_c^n\|\right)^{m(\theta^n)} \left(s_c^{n-1} + \tau_n \sqrt{\frac{2}{3}} \gamma_c^n\right)^{k(\theta^n)} \quad (5.4.24)$$

and rearrange it with respect to $\|\boldsymbol{\xi}_c^n\|$:

$$\|\boldsymbol{\xi}_c^n\| = \sqrt{\frac{2}{3}} \left(\frac{\gamma_c^n}{\sqrt{\frac{3}{2}} A(\theta^n) \left(s_c^{n-1} + \tau_n \sqrt{\frac{2}{3}} \gamma_c^n\right)^{k(\theta^n)}}\right)^{1/m(\theta^n)} . \quad (5.4.25)$$

Inserting (5.4.25) into (5.4.23), we finally obtain a scalar equation in which the creep multiplier $\gamma_c^n$ is the only unknown:

$$0 = \left| \sqrt{\frac{2}{3}} \left(\frac{\gamma_c^n}{A(\theta^n) \sqrt{\frac{3}{2}} \left(s_c^{n-1} + \tau_n \sqrt{\frac{2}{3}} \gamma_c^n\right)^{k(\theta^n)}}\right)^{\frac{1}{m(\theta^n)}} + \tau_n \gamma_c^n \left(2\mu(\theta^n, p^n) + \frac{2}{3} c_c(\theta^n) g(\gamma_c^n)\right)\right|$$

$$- \|\boldsymbol{\sigma}_t^{*n} - g(\gamma_c^n) \boldsymbol{X}_c^{n-1}\| .$$

$$(5.4.26)$$

This equation can be used for the determination of the creep multiplier $\gamma_c^n$. This can be realised by a numerical root finding algorithm, for instance by bisection method.

The resulting value of the creep multiplier $\gamma_c^n$ inserted in (5.4.23) yields the norm of the effective stress $\|\boldsymbol{\xi}_c^n\|$. After that, we can update the effective stress by (5.4.22), i.e.:

$$\boldsymbol{\xi}_c^n = \left(\boldsymbol{\sigma}_t^{*n} - g(\gamma_c^n) \boldsymbol{X}_c^{n-1}\right) \Big/ \left(1 + \tau_n \gamma_c^n \frac{1}{\|\boldsymbol{\xi}_c^n\|} \left(2\mu(\theta^n, p^n) + \frac{2}{3} c_c(\theta^n) g(\gamma_c^n)\right)\right) .$$

$$(5.4.27)$$

**Step 3: Update of all inelastic quantities, correction of stress tensor**

In the last step of our algorithm outlined in Box 5.4.1, we update the remaining quantities. We correct the deviator of the stress tensor $\boldsymbol{\sigma}^{*n}$ by equation (5.4.15):

$$\boldsymbol{\sigma}^{*n} = \boldsymbol{\sigma}_t^{*n} - 2\mu(\theta^n, p^n)\tau_n\gamma_c^n \frac{\boldsymbol{\xi}_c^n}{\|\boldsymbol{\xi}_c^n\|}, \tag{5.4.28}$$

and obtain the back stress $\boldsymbol{X}_c^n$ by (5.4.19),

$$\boldsymbol{X}_c^n = g(\gamma_c^n)\left(\boldsymbol{X}_c^{n-1} + \frac{2}{3}\tau_n c_c(\theta^n)\,\gamma_c^n \frac{\boldsymbol{\xi}_c^n}{\|\boldsymbol{\xi}_c^n\|}\right). \tag{5.4.29}$$

Finally, we update the creep strain

$$\boldsymbol{\varepsilon}_c^n = \boldsymbol{\varepsilon}_c^{n-1} + \tau_n\gamma_c^n \frac{\boldsymbol{\xi}_c^n}{\|\boldsymbol{\xi}_c^n\|}, \tag{5.4.30}$$

as well as the accumulated creep strain $s_c^n$:

$$s_c^n = s_c^{n-1} + \tau_n\sqrt{\frac{2}{3}}\,\gamma_c. \tag{5.4.31}$$

The corrected stress tensor follows by (5.4.11) using the updated value of $\boldsymbol{\varepsilon}_c^n$:

$$\begin{aligned}
\boldsymbol{\sigma}^n = {}& \boldsymbol{\sigma}_t^n - 2\mu(\theta^n, p^n)\tau_n\gamma_c^n \frac{\boldsymbol{\xi}_c^n}{\|\boldsymbol{\xi}_c^n\|} + \left(\lambda(\theta^n, p^n) + \frac{2}{3}\mu(\theta^n, p^n)\right)\operatorname{tr}(\boldsymbol{\varepsilon}^n)\boldsymbol{I} \\
& - K(\theta^n, p^n)\left(\frac{\varrho_0 - \varrho(\theta^n, p^n)}{\varrho(\theta^n, p^n)}\right)\boldsymbol{I},
\end{aligned} \tag{5.4.32}$$

which concludes our algorithm stated in Box 5.4.1.

### 5.4.3 Algorithm in the case of creep without back stress

In the case of creep without back stress, the evolution equation (5.4.1) simplifies to

$$\dot{\boldsymbol{\varepsilon}}_c = \frac{3}{2}A\left(\sqrt{\frac{3}{2}}\|\boldsymbol{\sigma}^*\|\right)^{m-1}\boldsymbol{\sigma}^*\, s_c^k. \tag{5.4.33}$$

The discretised version of the evolution of the creep strain is:

$$(\dot{\boldsymbol{\varepsilon}}_c)^n = \frac{3}{2}A_n\left(\sqrt{\frac{3}{2}}\|\boldsymbol{\sigma}^{*n}\|\right)^{m_n-1}\boldsymbol{\sigma}^{*n}(s_c^n)^{k_n} \tag{5.4.34}$$

where $A_n$, $m_n$ and $k_n$ represent the values of the material parameters at the current temperature $\theta^n$. We rearrange equation (5.4.34) and obtain

$$
\begin{aligned}
(\dot{\boldsymbol{\varepsilon}}_c)^n &= \sqrt{\frac{3}{2}}\, A_n \left( \sqrt{\frac{3}{2}} \|\boldsymbol{\sigma}^{*n}\| \right)^{m_n} (s_c^n)^{k_n} \frac{\boldsymbol{\sigma}^{*n}}{\|\boldsymbol{\sigma}^{*n}\|} \\
&= \gamma_c^n \frac{\boldsymbol{\sigma}^{*n}}{\|\boldsymbol{\sigma}^{*n}\|} ,
\end{aligned}
\tag{5.4.35}
$$

with

$$
\gamma_c^n := \sqrt{\frac{3}{2}}\, A_n \left( \sqrt{\frac{3}{2}} \|\boldsymbol{\sigma}^{*n}\| \right)^{m_n} (s_c^n)^{k_n} .
\tag{5.4.36}
$$

Considering the evolution of the accumulated creep strain, it holds that

$$
(\dot{s}_c)^n = \sqrt{\frac{2}{3}} \|(\dot{\boldsymbol{\varepsilon}}_c)^n\| = \sqrt{\frac{2}{3}} \gamma_c^n .
\tag{5.4.37}
$$

Summing up, we have to solve the equation

$$
(\dot{\boldsymbol{\varepsilon}}_c)^n = \gamma_c^n \frac{\boldsymbol{\sigma}^{*n}}{\|\boldsymbol{\sigma}^{*n}\|} .
\tag{5.4.38}
$$

As in Section 5.4.2, we will set up an equation which only depends on the creep multiplier $\gamma_c^n$. This equation will be solved numerically. After this, the current value of the creep strain follows by

$$
\boldsymbol{\varepsilon}_c^n = \boldsymbol{\varepsilon}_c^{n-1} + \tau_n \gamma_c^n \frac{\boldsymbol{\sigma}^{*n}}{\|\boldsymbol{\sigma}^{*n}\|} ,
\tag{5.4.39}
$$

using the approximation in (5.4.10).

Using (5.3.13) and (5.4.7), the current value of the deviator of the corrected stress tensor is given by

$$
\begin{aligned}
\boldsymbol{\sigma}^{*n} &= 2\mu_n(\boldsymbol{\varepsilon}^{*n} - \boldsymbol{\varepsilon}_c^n) \\
&= \boldsymbol{\sigma}_t^{*n} - 2\mu_n(\boldsymbol{\varepsilon}_c^n - \boldsymbol{\varepsilon}_c^{n-1}) .
\end{aligned}
\tag{5.4.40}
$$

And with (5.4.39), we obtain

$$
\boldsymbol{\sigma}^{*n} = \boldsymbol{\sigma}_t^{*n} - 2\mu_n \tau_n \gamma_c^n \frac{\boldsymbol{\sigma}^{*n}}{\|\boldsymbol{\sigma}^{*n}\|} .
\tag{5.4.41}
$$

In the next step, we calculate the required values in the following order:

$$
\gamma_c^n , \quad \boldsymbol{\sigma}^{*n} , \quad \boldsymbol{\varepsilon}_c^n , \quad s_c^n .
$$

First, we formulate the equation for the trial stress and have a look at the norm of the tensor. Rearranging equation (5.4.41) yields

$$
\boldsymbol{\sigma}_t^{*n} = \boldsymbol{\sigma}^{*n} + 2\mu_n \tau_n \gamma_c^n \frac{\boldsymbol{\sigma}^{*n}}{\|\boldsymbol{\sigma}^{*n}\|} .
\tag{5.4.42}
$$

Taking the norm of equation (5.4.42) we obtain

$$\|\boldsymbol{\sigma}_t^{*n}\| = \|\boldsymbol{\sigma}^{*n}\| \cdot \left| 1 + 2\mu_n \tau_n \gamma_c^n \cdot \frac{1}{\|\boldsymbol{\sigma}^{*n}\|} \right| . \tag{5.4.43}$$

As the expression in the absolute value in (5.4.43) is always positive, it follows that

$$\|\boldsymbol{\sigma}_t^{*n}\| = \|\boldsymbol{\sigma}^{*n}\| + 2\mu_n \tau_n \gamma_c^n . \tag{5.4.44}$$

Finally, we obtain the following equation for the norm of the deviatoric part of the corrected stress tensor, only depending on the scalar unknown $\gamma_c^n$:

$$\|\boldsymbol{\sigma}^{*n}\| = \|\boldsymbol{\sigma}_t^{*n}\| - 2\mu_n \tau_n \gamma_c^n , \tag{5.4.45}$$

As our aim is to solve the equation of the creep multiplier numerically, we need a second equation for the norm $\|\boldsymbol{\sigma}^{*n}\|$.

We use the definition of the creep multiplier (5.4.36) with equation (5.4.37),

$$\gamma_c^n = \sqrt{\frac{3}{2}} A_n \left( \sqrt{\frac{3}{2}} \|\boldsymbol{\sigma}^{*n}\| \right)^{m_n} \left( s_c^{n-1} + \tau_n \sqrt{\frac{2}{3}} \gamma_c^n \right)^{k_n} , \tag{5.4.46}$$

and rearrange it with respect to $\|\boldsymbol{\sigma}^{*n}\|$:

$$\|\boldsymbol{\sigma}^{*n}\| = \sqrt{\frac{2}{3}} \left( \frac{\gamma_c^n}{\sqrt{\frac{3}{2}} A_n \left( s_c^{n-1} + \tau_n \sqrt{\frac{2}{3}} \gamma_c^n \right)^{k_n}} \right)^{1/m_n} . \tag{5.4.47}$$

We consider the difference of the two equations (5.4.45) and (5.4.47) and obtain an equation which only depends on the scalar creep multiplier $\gamma_c^n$:

$$0 = \|\boldsymbol{\sigma}_t^{*n}\| - 2\mu_n \tau_n \gamma_c^n - \sqrt{\frac{2}{3}} \left( \frac{\gamma_c^n}{\sqrt{\frac{3}{2}} A_n \left( s_c^{n-1} + \tau_n \sqrt{\frac{2}{3}} \gamma_c^n \right)^{k_n}} \right)^{1/m_n} . \tag{5.4.48}$$

Thus, we reduced our three-dimensional problem to a one-dimensional problem which can be solved numerically with respect to $\gamma_c^n$. This can be realised by a numerical root finding algorithm, for instance by bisection method.

**Step 3: Update of all inelastic quantities, correction of stress tensor**

Having determined the creep multiplier $\gamma_c^n$, we can update all other inelastic quantities.

It can easily be shown that it holds

$$\frac{\boldsymbol{\sigma}^{*n}}{\|\boldsymbol{\sigma}^{*n}\|} = \frac{\boldsymbol{\sigma}_t^{*n}}{\|\boldsymbol{\sigma}_t^{*n}\|} . \tag{5.4.49}$$

Taking this into account and using equation (5.4.41), the updated deviator of the stress tensor follows by

$$\boldsymbol{\sigma}^{*n} = \left( 1 - \frac{2\mu_n \tau_n \gamma_c^n}{\|\boldsymbol{\sigma}_t^{*n}\|} \right) \boldsymbol{\sigma}_t^{*n} \, . \tag{5.4.50}$$

Now, we are able to determine the current values of the creep strain $\varepsilon_c^n$ and of the accumulated creep strain $s_c^n$ with

$$\boldsymbol{\varepsilon}_c^n = \boldsymbol{\varepsilon}_c^{n-1} + \tau_n \gamma_c^n \frac{\boldsymbol{\sigma}^{*n}}{\|\boldsymbol{\sigma}^{*n}\|} \, , \tag{5.4.51}$$

and

$$s_c^n = s_c^{n-1} + \tau_n \sqrt{\frac{2}{3}} \gamma_c^n \, . \tag{5.4.52}$$

Finally, we correct the stress tensor with the updated value of $\boldsymbol{\varepsilon}_c$ using (5.4.11):

$$\boldsymbol{\sigma}^n = \boldsymbol{\sigma}_t^n - 2\mu(\theta^n, p^n)\tau_n \gamma_c^n \frac{\boldsymbol{\sigma}_t^{*n}}{\|\boldsymbol{\sigma}_t^{*n}\|} + (\lambda(\theta^n, p^n) + \frac{2}{3}\mu(\theta^n, p^n))\mathrm{tr}(\boldsymbol{\varepsilon}^n)\boldsymbol{I}$$
$$- K(\theta^n, p^n) \left( \frac{\varrho_0 - \varrho(\theta^n, p^n)}{\varrho(\theta^n, p^n)} \right) \boldsymbol{I} \, . \tag{5.4.53}$$

# 6 Simulation results, validation with experimental data

In this chapter, we present results of simulations using the implemented model equations given in Chapter 2 and the algorithms developed in Chapters 4 and 5. Furthermore, we provide experimental data as well as results from parameter identification.

Here, we distinguish between the verification and the validation of a model following the philosophy of Mahnken (2004). For the verification of a model, one-dimensional experimental data is used. The necessary data are frequently obtained by uniaxial experiments with special testing devices. By comparing the model response with the data, it is checked whether the model is generally capable to describe the material behaviour. This is used to identify material parameters. In the next step, the material parameters thus obtained are used for general 3D simulations. The comparison between 3D workpiece experiments and 3D simulations is called validation. In the following, we also refer to these steps as 'direct problem' when assuming certain material laws and as 'inverse problem' corresponding to parameter identification.

In particular, Section 6.1 presents creep experiments under constant stress and temperature, Section 6.2 shows data stemming from austenitisation experiments under stress.

We apply the numerical algorithm for the verification of creep and TRIP behaviour presented in Chapter 4. We use these models together with the experimental data presented here to identify the required material parameters. In order to be able to describe creep during heating, our aim is to identify parameter functions depending on temperature.

The obtained parameters for the creep strains are used in the 3D model which enables us to implement the model and to perform simulations. The results are presented in Section 6.3. Here, the material behaviour of different workpieces during certain heat treatment scenarios is investigated. Moreover, we present experimental data stemming from workpiece experiments. By comparing our simulated data to the experimental results, we are able to validate our model, see Section 6.4.

Further details about the uniaxial experiments as well as about the parameter identification can be found in Wolff et al. (2013) and Bökenheide et al. (2012a). The presented experimental data was provided by the Institute of Materials Science Bremen (IWT).

Figure 6.1: Gleeble® testing device with steel specimen.

# 6.1 Creep of the bearing steel SAE 52100 (100Cr6) (1D case)

In the following, we present data from uniaxial experiments using small steel specimen. The performed experiments aimed to examine creep behaviour of the steel SAE 52100 (100Cr6) under different fixed tempratures and stresses. Details about the chemical composition of the investigated steel are given in Acht et al. (2008a) and Acht et al. (2008b).

The experimental data is used for the verification of material laws and to determine certain material parameters. Section 6.1.2 provides results.

More details about experimental and simulation results can be found in Wolff et al. (2012c). In Bökenheide et al. (2011) and Bökenheide et al. (2012b), different creep models are compared. The authors' investigations are based on the same experimental data as presented here.

## 6.1.1 Results from uniaxial experiments

The uniaxial experiments presented in the following were performed in two experimental sets: one using the initial state of the material, the second with the austenitised material.

The experimental data was obtained with a servo-hydraulic testing device of type Gleeble®. A picture of the steel specimen positioned in the testing device can be seen in Figure 6.1. For the geometry of the specimen and further technical details, see Dalgic et al. (2009).

The creep experiments were performed under certain fixed 'nominal' stresses and temperatures. Figure 6.2 gives an overview of the different experiments.

By means of the measured change in length in longitudinal and transversal direction, one obtains the experimental strains $\epsilon_L$ and $\epsilon_D$.

We obtain the *experimental* creep strain $\epsilon_{c,exp}$ (cf. Section 4.2) via the formula

$$\epsilon_{c,exp}(t) = \frac{2}{3}(\epsilon_L(t) - \epsilon_D(t)) - \frac{S(t)}{3\mu(\theta(t), p(t))} \ . \tag{6.1.1}$$

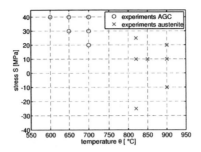

Figure 6.2: Overview of performed creep experiments: Temperature and stresses of experiments using AGC (**o**) as well as austenite (**x**).

### Creep of the initial material

The initial state of 100Cr6 steel consists of a ferrite matrix with spheroidized carbide inclusions stemming from annealing on globular cementite. In the following, we abbreviate this state as AGC.

Table 6.1 summarises the fixed stresses and temperatures during the experiments using AGC. The measured longitudinal and transversal strains are summarised in Figure 6.3. The resulting creep curves using formula (6.1.1) are plotted in Figure 6.4.

| $\theta[°C]$ | 600 | 650 | 650 | 700 | 700 |
|---|---|---|---|---|---|
| $S[\text{MPa}]$ | 40 | 30 | 40 | 20 | 30 |

Table 6.1: Creep experiments: initial material (AGC).

### Creep of austenite

During high temperatures, the initial material transforms into austenite. In order to be able to describe creep during heating, we consider a second experimental set using the austenitised material. Again, each is performed under fixed nominal stresses and temperatures. An overview of the performed experiments using austenite is given in Table 6.2.

In the same way as described above, we obtain the experimental creep strains for austenite, using formula (6.1.1). The results are shown in Figure 6.6. In Figure 6.5, the corresponding longitudinal and transversal strains are plotted.

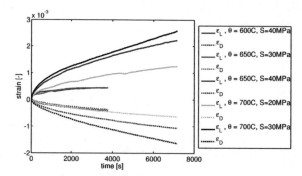

Figure 6.3: Creep tests using AGC under different fixed temperatures $\theta$ and stresses $S$: experimental longitudinal (solid line) and transversal (dashed line) strains $\epsilon_L$ and $\epsilon_D$, respectively.

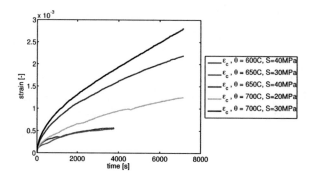

Figure 6.4: Creep tests using AGC under different fixed temperatures $\theta$ and stresses $S$: experimental creep strains $\epsilon_c$.

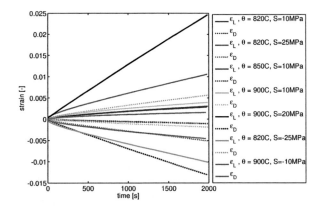

Figure 6.5: Creep tests using austenite under different fixed temperatures $\theta$ and stresses $S$: longitudinal (solid line) and transversal (dashed line) strains $\epsilon_L$ and $\epsilon_D$, respectively.

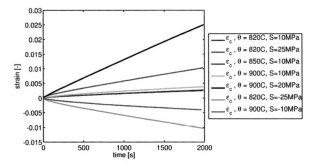

Figure 6.6: Creep tests using austenite under different fixed temperatures $\theta$ and stresses $S$: experimental creep strains $\epsilon_c$.

| $\theta[°C]$ | 820 | 820 | 850 | 900 | 900 | 820 | 900 |
|---|---|---|---|---|---|---|---|
| $S[MPa]$ | 10 | 25 | 10 | 10 | 20 | -25 | -10 |

Table 6.2: Creep experiments: austenite.

| Notation | Model | $k$ | $l$ |
|---|---|---|---|
| **M1** | creep without backstress, (4.1.8) | $k < 0$ | – |
| **M2** | Armstrong-Frederick, (4.1.8),(4.1.10) | $k < 0$ | $l = 1$ |
| **M3** | Robinson, (4.1.8),(4.1.10) | $k < 0$ | $l = 0$ |

Table 6.3: Overview of models.

## 6.1.2 Results from parameter identifications

The optimisation procedure presented in Chapter 4 in Section 4.6 is applied to the concrete material behaviour of 100Cr6 (SAE 52100) steel. Our aim is to determine the material parameters of the model equations (4.1.8), (4.1.10). We use the experimental data presented above (see Tables 6.1 and 6.2).

The procedure involves both the strain- and the stress-driven approach. As generally the presented method works also for non-constant stress, the calculations were performed using the measured stress that is usually varying slightly. First, we focus on the material behaviour of the initial state. After that, we determine the parameters for austenite. The obtained parameters are used to calculate the creep strain and to compare it to the experimental data. The results for both states of the material are presented in the following.

Regarding the model (4.1.8), (4.1.10), we use the following specifications:

- *we assume primary creep, thus $k < 0$*

- *a constant drag stress $D_c = 1 MPa$ is applied*

- *two model approaches are considered (cf. Table 6.3):*
  - *(i) creep without back stress, thus $A, m, k$ in (4.1.8) have to be determined (model M1)*
  - *(ii) creep with back stress and a back stress evolution according to (4.1.10) where $l = 1$ (Armstrong-Frederick type, model M2). The parameters $A, m, k$ as well as $b_c, c_c$ have to be determined.*

The parameters sought are set as parameter functions depending on temperature. This enables us to handle creep under varying temperatures. Based on experience and former investigations, the temperature dependencies of the material parameters are chosen as:

$$A(\theta) = A_0 \exp\left(\frac{-Q}{R\,(273.15 + \theta)}\right), \qquad m(\theta) = m_0 + m_1\theta, \qquad k(\theta) = k_0 + k_1\theta,$$

$$c_c(\theta) = c_0 + c_1\theta, \qquad\qquad\qquad b_c(\theta) = b_0 + b_1\theta, \qquad\qquad (6.1.2)$$

where $Q$ is the activation energy and $R = 8.314\,\frac{J}{mol\,K}$ states the gas constant (for convenience, here we use degrees Centigrade). Altogether, the following material parameters have to be determined:

$$A_0, \qquad Q, \qquad m_0, \qquad m_1, \qquad k_0, \qquad k_1, \qquad c_0, \qquad c_1, \qquad b_0, \qquad b_1,$$

for both AGC and austenite.

Our aim is to find the parameter set that minimises the sum in (4.6.5). Regarding the minimisation, the MATLAB® routines `fminsearch` and `fmincon` are used (cf. Remark 6.1.1).

We aim to find parameters which represent the best fit to the given experimental data under the consideration of both the strain- and the stress-driven approach. Using the parameters obtained in this way, the creep strain can be calculated with the strain-driven as well as with the stress-driven approach. Figures 6.7 – 6.12 show some results.

In the following Section 6.2.2, the obtained (temperature-depending) material parameters for each phase will be used to model th simultaneous occurrence of creep and TRIP during heating and austenitisation.

For further details we refer to Wolff et al. (2012c), Bökenheide et al. (2011) and Bökenheide et al. (2012b).

**Remark 6.1.1** (Optimisation under constraints). *In addition to the procedure stated above, we consider a second optimisation procedure. Here, the optimisation was conducted for model (4.1.8) without back stress, where only the strain-driven approach is taken into account. For the minimisation of the cost functional, the MATLAB® routine* `fmincon` *is used. The function takes the following constraints for the parameters into account*

$$A > 0 \quad , \quad m > 0 \quad , \quad k < 0, \qquad\qquad (6.1.3)$$

*which must be fulfilled in the considered temperature interval of the corresponding phase.*

**Creep of the initial phase**

The parameters obtained by the identification procedure over *all* AGC creep experiments are (see Remark 6.1.2):

- For model M1, i.e. for model (4.1.8) without back stress,

$$A_0 = 7.469 \cdot 10^{-9} \, s^{-1}, \quad Q = 1.903 \cdot 10^5 \, J(mol)^{-1}, \quad m_0 = 11.668,$$
$$m_1 = -0.0098 \, K^{-1}, \quad k_0 = 2.651, \quad k_1 = -0.0061 \, K^{-1},$$

$$(6.1.4)$$

- for model M2, i.e. for model (4.1.8), (4.1.10) with $l = 1$,

$$A_0 = 1.5374 \cdot 10^{-11} \, s^{-1}, \quad Q = 1.3998 \cdot 10^5 \, J(mol)^{-1}, \quad m_0 = 9.3184,$$
$$m_1 = -0.0064 \, K^{-1}, \quad k_0 = 2.159, \quad k_1 = -0.0054 \, K^{-1},$$
$$c_0 = 6.5638 \cdot 10^{-5} \, MPa, \quad c_1 = 0.027 \, MPa \, K^{-1},$$
$$b_0 = -9.9792 \cdot 10^{-6}, \quad b_1 = 0.0034 \, K^{-1}.$$

$$(6.1.5)$$

The parameters are used to calculate the creep strain by the strain-driven as well as by the stress-driven approach. The results are presented in Figures 6.7 and 6.8.

Here, the creep strain of the initial state is described quite well regarding the strain-driven approach. The creep curves of AGC calculated by the stress-driven mode differ from experiments. A cause for this may be the composite structure of AGC requiring a more complex creep model.

As the results for AGC are deviating from the experimental data, a second parameter identification is considered. The experimental data of AGC was cut differently and the optimisation was conducted under constraints (see Remark 6.1.1). Here, the procedure took only the strain-driven approach into account as it showed better results in the first attempt. The parameter function with the start and optimal values are illustrated in Figure 6.9. The obtained parameters are:

$$A_0 = 0.51066 \, s^{-1}, \quad Q = 1.90255 \cdot 10^5 \, J(mol)^{-1}, \quad m_0 = 11.50225,$$
$$m_1 = -0.00515 \, K^{-1}, \quad k_0 = 2.4528, \quad k_1 = -0.00885 \, K^{-1}, \quad (6.1.6)$$

The results are shown in Figure 6.10.

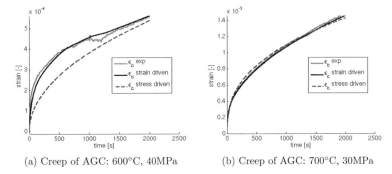

(a) Creep of AGC: 600°C, 40MPa    (b) Creep of AGC: 700°C, 30MPa

Figure 6.7: Creep of AGC: Experiment and simulation with model (4.1.8) without back stress using strain- and stress-driven driven approach.

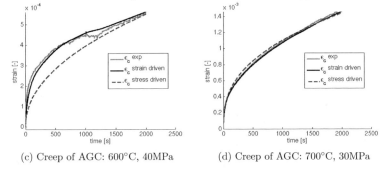

(c) Creep of AGC: 600°C, 40MPa    (d) Creep of AGC: 700°C, 30MPa

Figure 6.8: Creep of AGC: Experiment and simulation using model (4.1.8), (4.1.10) with $l = 1$ (model by Armstrong Frederick) with strain- and stress-driven approach.

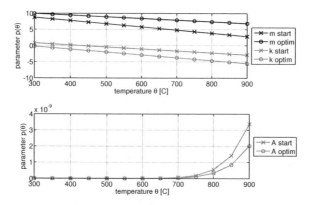

Figure 6.9: Results from parameter identification: start ($\mathbf{x}$) and optimal ($\mathbf{o}$) values (AGC).

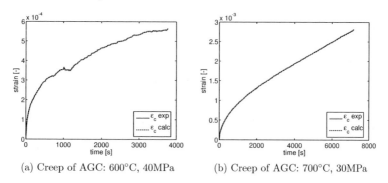

(a) Creep of AGC: 600°C, 40MPa          (b) Creep of AGC: 700°C, 30MPa

Figure 6.10: Creep of AGC: Experiment and simulation with model (4.1.8) without back stress using strain-driven approach.

**Creep of austenite**

Analogously, we proceed with the creep experiments of austenite. The corresponding parameters for austenite obtained by the identification procedure over *all* creep experiments are:

- For model M1, i.e. (4.1.8) without back stress

$$A_0 = 1.628 \cdot 10^{-9}\, s^{-1}\,, \quad Q = 1.441 \cdot 10^4\, J(mol)^{-1}\,, \quad m_0 = -4.027\,,$$
$$m_1 = 0.0078\, K^{-1}\,, \quad\quad k_0 = -1.29\,, \quad\quad\quad\quad k_1 = 0.001\, K^{-1}\,.$$
$$(6.1.7)$$

- For model M2, i.e. (4.1.8), (4.1.10) with $l = 1$,

$$A_0 = 5.591 \cdot 10^{-10}\, s^{-1}\,, \quad Q = 1.183 \cdot 10^4\, J(mol)^{-1}\,, \quad m_0 = -8.352\,,$$
$$m_1 = 0.013\, K^{-1}\,, \quad\quad k_0 = -3.547\,, \quad\quad\quad\quad k_1 = 0.0037\, K^{-1}\,,$$
$$c_0 = 8.22 \cdot 10^{-10}\, \mathrm{MPa}\,, \quad c_1 = 0.013\, \mathrm{MPa}\, K^{-1}\,,$$
$$b_0 = 1.395 \cdot 10^{-7}\,, \quad\quad b_1 = 0.015\, K^{-1}\,.$$
$$(6.1.8)$$

The resulting creep strains are given in Figures 6.11 and 6.12. The simulated creep strains of austenite show a good agreement with the experimental data for both the strain-driven and the stress-driven approach.

**Remark 6.1.2** (Comments on optimisation procedure).

(i) *Identification procedure 'over all AGC creep experiments', e.g., means that the optimisation procedure uses the data of all AGC creep experiments simultaneously. Therefore, the cost functional is actually a sum of cost functionals which involve the single data sets (see (4.6.5)). Thus, the parameters in (6.1.4) - (6.1.8) are optimal with respect to the* whole *data set of the corresponding material and not to a single experiment (cf. Mahnken and Stein (1996) for discussion and further references).*

(ii) *Before the identification of the parameters, first of all, a smoothing of the experimental data is performed. This is realised by the MATLAB® routine* filter.
*As numerical optimisations only return local minima depending on the choice of the initial values, a set of several different starting values for each parameter is used. We perform several optimisations using all combinations of initial values, which yields the optimal parameter set.*

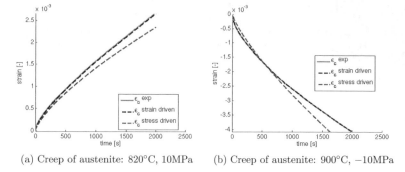

(a) Creep of austenite: 820°C, 10MPa     (b) Creep of austenite: 900°C, −10MPa

Figure 6.11: Creep of austenite: Experiment and simulation using model (4.1.8) without back stress with strain- and stress-driven approach.

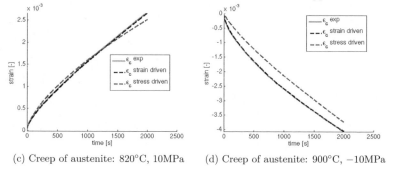

(c) Creep of austenite: 820°C, 10MPa     (d) Creep of austenite: 900°C, −10MPa

Figure 6.12: Creep of austenite: Experiment and simulation using model (4.1.8), (4.1.10) with $l = 1$ (model by Armstrong Frederick) with strain- and stress-driven approach.

## 6.2 Creep and TRIP of the bearing steel SAE 52100 (100Cr6) (1D case)

Next, we consider creep during non-isothermal experiments. We intend to model the material behaviour during heating and austenitisation which includes creep, TRIP and phase transformations. Therefore, we determine the material parameters involved in the material law for TRIP (4.1.11) by a parameter identification using uniaxial experimental data.

During austenitisation under stress, creep and TRIP occur together. Thus, the (longitudinal) inelastic strain is the sum

$$\epsilon_{in} = \epsilon_c + \epsilon_{trip} \ . \tag{6.2.1}$$

| $S$[MPa] | 0 | 5 | 10 | 15 | 20 | -5 | -10 | -15 | -20 | -25 |
|---|---|---|---|---|---|---|---|---|---|---|

Table 6.4: Heating experiments.

In order to determine the TRIP strain, we use the experimental inelastic strain and the creep strain of the phase mixture obtained by means of the parameters determined beforehand.

First, we present data from austenitisation experiments conducted under different fixed stresses. After that, we focus on the arising phase transformations. The phase fractions are used to determine the creep strain of the phase mixture calculated by a mixture rule. Then, the difference $\epsilon_{in} - \epsilon_c$ can be compared to $\epsilon_{trip}$ in order to identify the required material parameters, see Section 6.2.2.

### 6.2.1 Experimental results

The experiments that are used in the subsequent investigations, were performed under different fixed 'nominal' stresses under the same temperature course. An overview of the given stresses during the experiments is given in Table 6.4.

Figure 6.13 presents the measured longitudinal and transversal strains $\epsilon_L$ and $\epsilon_D$. These are used in order to obtain the experimental inelastic strain. The results are shown in Figure 6.14.

### 6.2.2 Simulation results

First, we focus on the arising phase transformations. After that, we determine the parameters which are needed for the calculation of the TRIP strain.

**Phase fractions during austenitisation**

During heating, the initial material (AGC) transforms into austenite. We determine the corresponding phase fractions, i.e. ferrite, carbide and austenite fractions, via the extended model by Johnson-Mehl-Avrami-Kolmogoroff which was presented in Chapter 2, see equations (2.3.3)-(2.3.4). The values $p_{0,F}$, $p_{0,C}$, $p_{eq,F}(\theta)$, $p_{eq,C}(\theta)$ have been provided by the Institute of Materials Science Bremen (IWT). Following Reichelt (1982), a value of $k_c = 10.3$ has been used.

We follow Surm et al. (2008), where a distinction between the transformations during heating up to the holding temperature (here, 850°C) and during holding is suggested.

During heating, the parameters $\tau_F, \tau_C$ are modelled as

$$\tau_i(\theta, \nu) = \tau_{0,i}(\nu) \exp\left(\frac{-Q}{R\,\theta_{abs}}\right), \qquad (6.2.2)$$

where $i$ stands for $F$ and $C$, respectively, $Q$ denotes the activation energy for carbon diffusion in austenite, $R$ stands for the gas constant and $\theta_{abs}$ denotes the

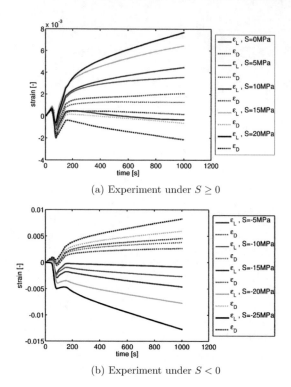

(a) Experiment under $S \geq 0$

(b) Experiment under $S < 0$

Figure 6.13: Heating experiments: Longitudinal and transversal strains $\epsilon_L$, $\epsilon_D$.

absolute temperature in Kelvin. During holding, the ferrite phase is assumed to have been completely transformed. Thus, the following parameters have to be determined: $k_C$, $\tau_{0,F}(\nu)$, $n_F$, $\tau_{0,F}(\nu)$, $n_C$ during heating and $\tau_C$, $n_C$ during holding.

By means of the bulk density $\rho$ at current temperature $\theta(t)$ and phase fractions $p(t)$ which can be expressed by the mixture rule

$$\varrho(\theta, p) = \varrho_F(\theta)\, p_F + \varrho_C(\theta)\, p_C + \varrho_A(\theta, u_c)\, p_A \,, \qquad (6.2.3)$$

(see (2.3.6)), we obtain the part of the volume strain caused by density changes (see Wolff et al. (2012c)):

$$\epsilon_{V,\rho}(t) := 3\left(\sqrt[3]{\frac{\rho_0}{\rho(\theta(t), p(t))}}\right) \,. \qquad (6.2.4)$$

Using formula (4.2.5) for $\epsilon_V$, this can be compared with

$$\epsilon_L(t) + 2\,\epsilon_D(t) - \frac{(1 - 2\,\nu(\theta(t), p(t)))}{E(\theta(t), p(t))}\, S(t) \,. \qquad (6.2.5)$$

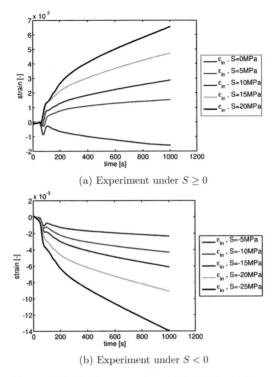

(a) Experiment under $S \geq 0$

(b) Experiment under $S < 0$

Figure 6.14: Heating experiments: Inelastic strain $\epsilon_{in}$.

The expression in (6.2.5) is calculated by the given experimental data. The material parameters involved in the equations (2.3.3)-(2.3.4) can thus be determined via a least-square fit.

As a result, we obtain fitted evolution curves for $p_F$, $p_C$ and $p_A$ for each experiment. Figure 6.15 shows an example of the resulting calculated phase fractions of ferrite, carbide and austenite during heating from 700°C to 850°C together with the respective temperature (cf. Remark 6.2.1).

After this, we are able to determine the creep strain of the phase mixture using the mixture rule

$$\varepsilon_c(t) = (p_F(t) + p_C(t))\, \varepsilon_{c,AGC}(t) + p_A(t)\, \varepsilon_{c,A}(t) . \qquad (6.2.6)$$

Here, $\varepsilon_{c,AGC}$ and $\varepsilon_{c,A}$ denote the creep strains of the initial material (AGC) and of austenite, respectively, which are calculated using the temperature-dependent material parameters obtained beforehand, see Section 6.1.2.

**Remark 6.2.1.** *Generally, phase transformations are stress-dependent. However, in the available experimental data no clear tendency could be detected. Therefore, this dependency is dropped in the modelling.*

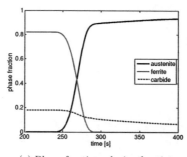

(a) Phase fractions during heating.

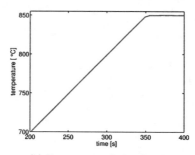

(b) Temperature during heating.

Figure 6.15: Creep and TRIP during austenitisation: Phase fractions and temperature during heating under stress ($S = 10$MPa).

### Creep and transformation-induced plasticity

During austenitisation under stress, creep and TRIP occur together. Thus, the inelastic strain $\epsilon_{in}$ corresponds to the sum:

$$\epsilon_{in}(t) = \epsilon_c(t) + \epsilon_{trip}(t) \ . \tag{6.2.7}$$

The inelastic strain $\epsilon_{in}$ can be obtained by the experimental data via formula (4.2.9). The creep strain of the phase mixture can be calculated by means of (6.2.6) and can thus be assumed as known, too. Hence, the TRIP strain can be obtained from experimental data by (6.2.7).

Now, assuming the material law for $\epsilon_{trip}$,

$$\dot{\epsilon}_{trip}(t) = \kappa(\theta(t), S(t))\, S(t)\, \frac{\mathrm{d}\phi(p(t))}{\mathrm{d}p(t)}\dot{p}(t)\,, \tag{6.2.8}$$

with the simple saturation function $\phi(p) := p$, the material parameter $\kappa$ is determined via the identification procedure developed in Chapter 4, see Sections 4.5 and 4.6.

Here, we choose the parameter $\kappa$ to be linearly dependent on temperature, i.e.

$$\kappa(\theta) = \kappa_0 + \kappa_1\,\theta\,. \tag{6.2.9}$$

Therefore, $\kappa_0$ and $\kappa_1$ must be determined. First, a fitting for each experiment is performed individually. The resulting values for $\kappa_0, \kappa_1$ are given in Table 6.5. The values obtained via fitting over all data are (see Remark 6.1.2):

$$\kappa_0 = 1.223 \cdot 10^{-4}\,\mathrm{MPa}^{-1}\,, \qquad \kappa_1 = 1.165 \cdot 10^{-7}\,\mathrm{MPa}^{-1}\mathrm{K}^{-1}\,. \tag{6.2.10}$$

Figure 6.16 shows the TRIP strain calculated with parameters obtained via 'individual' optimisation ($\epsilon_{trip\_ind}$) and via optimisation over all data ($\epsilon_{trip\_sum}$) together with the difference $\epsilon_{in} - \epsilon_c$ in the respective experiment. Here, the creep strain is calculated with model M2 (left) as well as with model M1 (right) (see Table 6.3). The TRIP strain is calculated by (6.2.8).

Qualitatively, the model is capable to describe the material behaviour. However, contrary to the simulated creep strain curves in Section 6.1.2, there are considerable deviations between $\epsilon_{trip,exp} = \epsilon_{in} - \epsilon_c$ and $\epsilon_{trip,calc}$. Relevant causes for this may be the calculation of the creep strain via a linear mixture rule and an extrapolation of the creep behaviour of AGC and austenite outside of their original temperature intervals (cf. Figure 6.2).

**Remark 6.2.2.**     *(i) The ferrite and carbide fractions tend asymptotically to zero. Thus, in the context of equations (2.3.3) - (2.3.4), 'completely' means that the remaining ferritic phase fraction at time $t = t_A$ is very small (e.g. by setting $p_F(t_A) = 0.01$ or $p_F(t_A) = 0.001$).*

*(ii) The initial values $p_{0,F}$ and $p_{0,C}$ in the phase model (2.3.3)-(2.3.4) must be determined beforehand. The equilibrium values $p_{eq,F}$ and $p_{eq,C}$ can be used from literature (cf. Surm et al. (2008)).*

*(iii) Concerning the approach behind the equations (2.3.3) - (2.3.4) of the phases' dissolution, we refer to Wolff et al. (2007), e.g., for further discussion.*

**Conclusion to parameter identification**     Altogether, the results of the model verification for the 1D case show that the introduced models are in accordance with real experimental data. The verification of the creep models yield good results for both the initial state AGC as well as for austenite.

|        | $S = 5$           | $S = 10$          | $S = 15$          | $S = 20$          | $S = -5$          | $S = -15$         |
|--------|-------------------|-------------------|-------------------|-------------------|-------------------|-------------------|
| $\kappa_0$ | $1.22{\cdot}10^{-4}$ | $1.22{\cdot}10^{-4}$ | $1.21{\cdot}10^{-4}$ | $1.22{\cdot}10^{-4}$ | $1.21{\cdot}10^{-4}$ | $1.23{\cdot}10^{-4}$ |
| $\kappa_1$ | $1.56{\cdot}10^{-7}$ | $4.11{\cdot}10^{-8}$ | $5.9{\cdot}10^{-8}$ | $2.65{\cdot}10^{-8}$ | $1.47{\cdot}10^{-7}$ | $1.18{\cdot}10^{-7}$ |
|        | $S = -20$         | $S = -25$         |                   |                   |                   |                   |
| $\kappa_0$ | $1.21{\cdot}10^{-4}$ | $1.21{\cdot}10^{-4}$ |                   |                   |                   |                   |
| $\kappa_1$ | $1.47{\cdot}10^{-7}$ | $1.99{\cdot}10^{-7}$ |                   |                   |                   |                   |

Table 6.5: Values of $\kappa_0\,[(\mathrm{MPa})^{-1}]$ and $\kappa_1\,[(\mathrm{MPa})^{-1}\mathrm{K}^{-1}]$ corresponding to experiments under different stresses $S\,[\mathrm{MPa}]$.

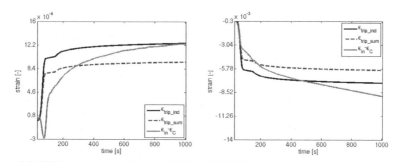

(a) TRIP during austenitisation: 5MPa    (b) TRIP during austenitisation: $-25$MPa

Figure 6.16: TRIP strain during austenitisation: Experiment (grey) and simulation.

Considering the investigation of austenitisation experiments under stress, the application of the developed algorithms yields reasonable results. However, in contrast to the results of simulating the creep strain as a single phenomenon, the simultaneous occurrence of creep and TRIP represents a special challenge. The inelastic strain of the phase mixture is assumed as the sum of the TRIP strain and the creep strain. The latter must be calculated using a mixture rule and extrapolation which might be a source of errors. Moreover, the calculated bulk creep strain depends on the used model, and thus influences the calculated TRIP strain.

## 6.3 3D Simulation and validation with workpiece experiments

The background of our investigations is the material behaviour of steel during heat treatment experiments with steel workpieces. Figure 6.19 shows the experimental setup of a steel ring which is heated and quenched. The ring is measured afterwards where one examines a deformation of the workpiece. The reason for this is assumed to be transformation-induced plasticity and classical plasticity which arise during the quenching of the ring. In the performed simulations of quenching there were deviations between the simulation and the experimental results (cf. Suhr (2010)). Therefore, it is possible that further effects arising during the heating phase of the experiment lead to a permanent deformation of the workpiece. As the steel ring is positioned on a support during heat treatment, inner stresses due to bending are possible. These stresses can lead to inelastic effects during heating, i.e. creep.

Unfortunately, it is not possible to measure these specific effects *directly* in the mentioned experimental setup. The workpiece is located in the furnace at this time of the experiment and is quenched afterwards. However, we can deduce the impact of this effect on the magnitude of the deformation of the workpiece. Therefore, we performed 3D simulations of the workpiece, enabling us to investigate the material behaviour during heating. The results are presented in Section 6.4.4.

Furthermore, we will consider experiments using a steel beam which were performed in order to observe particularly the single effect of creep on the workpiece. The experimental setup is as follows: The steel beam is located on each of its end on a support, see experimental setup in Figure 6.18. During the experiment, the beam is heated up in the furnace to a specific temperature. After this, the beam cools down to room temperature. During the experiment, there arise a bending of the workpiece which leads to inner stresses and thus to creep. The resulting permanent deformation of the beam is measured afterwards.

By means of the measured deformation, we are able to compare simulated data with the experimental data of the cooled beam. We performed simulations with the implemented geometry of the beam by giving the same external temperature course as during the experiment. The results will be presented in Section 6.4.3.

The 3D simulations presented in the following were performed with the open source Finite Element Toolbox ALBERTA (see logo in Figure 6.17), details can be found in Schmidt and Siebert (2005). As this software is open source, there are less restrictions than in commercial Finite Element tools and gives us possibility to implement further models.

The results were illustrated by means of the open source scientific visualisation ParaView.

Figure 6.17: Logo Finite Element Toolbox ALBERTA.

## 6.4 Creep during heating and austenitisation (3D case)

We consider two different experimental setups using different workpieces. First, we present creep experiments using a steel beam that is located on a support on each of its end. Section 6.4.1 presents the 3D simulation results. These are verified by means of experimental data, see Section 6.4.3.

After that, we will consider experiments using a steel ring which is heated and quenched. The simulation results obtained with the implemented geometry of the ring will be presented in Section 6.4.4.

Figure 6.18: Creep experiment: Steel beams in front of furnace[1].

Figure 6.19: Heat treatment experiment: Steel ring in gas nozzle field[1].

### 6.4.1 3D Simulation of a beam

Figure 6.20 shows the implemented geometry of the beam. The size of the beam in $x, y, z$-directions (according to the axes in Figure 6.20) is: $l_x = 0.5\,m$, $l_y = 0.03\,m$, $l_z = 0.02\,m$.

---

[1]Pictures provided by Institute of Materials Science Bremen (IWT).

In Figure 6.21, the effect of gravity is visualised. The figure shows the deformed beam (coloured beam) together with the undeformed geometry (grey beam). The defor

Figure 6.20: Simulated beam with mesh.

Figure 6.21: Deformation at $t = 7950s$, $\theta = \text{RT}$ (after heating to $\theta = 700°\text{C}$, see Figure 6.28): undeformed geometry (grey) and deformed beam (deformation graphically increased by factor $2 \cdot 10^2$).

Figure 6.22 shows the heat propagation in the inside of the beam. The figure presents a horizontal cut of the simulated workpiece.

In the following, we present some results of different simulations. To be more specific, we consider the simulation results of the material behaviour during:

- heating
- heating and austenitisation
- heating and cooling
- heating, austenitisation and cooling.

| Notation | $\theta_0$ | $\theta_{max}$ | $\theta_{end}$ |
|----------|------------|----------------|----------------|
| **S1**   | $RT$       | 600            | 600            |
| **S2**   | $RT$       | 850            | 850            |
| **S3**   | 600        | 600            | 600            |

| Notation | $\theta_0$ | $\theta_{max}$ | $\theta_{end}$ |
|----------|------------|----------------|----------------|
| **S4**   | $RT$       | 700            | $RT$           |
| **S5**   | $RT$       | 940            | $RT$           |
| **S6**   | 600        | 940            | $RT$           |

Table 6.6: Overview of performed simulations without (left) and with cooling (right).

| Experiments | $\theta_0$ | $\theta_{max}$ | $\theta_{end}$ |
|-------------|------------|----------------|----------------|
| **E1,E2**   | $RT$       | 700            | $RT$           |
| **E3,E4**   | $RT$       | 940            | $RT$           |

Table 6.7: Overview of performed experiments.

Considering the last two points, we especially focus on the deformation of the workpiece at the end of the simulation. Table 6.6 summarises the initial, maximal and final temperatures during the different simulations.

In the following, the entries of the creep strain tensor are denoted as follows:

$$
\varepsilon_c = \begin{pmatrix} \varepsilon_{cxx} & \varepsilon_{cxy} & \varepsilon_{cxz} \\ \varepsilon_{cxy} & \varepsilon_{cyy} & \varepsilon_{cyz} \\ \varepsilon_{cxz} & \varepsilon_{cyz} & -\varepsilon_{cxx} - \varepsilon_{cyy} \end{pmatrix} . \tag{6.4.1}
$$

Furthermore, the displacement vector $\boldsymbol{u}$ is denoted as $\boldsymbol{u} = (u_x, u_y, u_z)$.

**Creep during heating (S1)**  The given external temperature in this simulation increases to $\theta_{max} = 600°C$ and is kept constant afterwards (see Figure 6.23). Figure 6.24 shows the resulting diagonal entries of the creep strain tensor $\varepsilon_c$ in the last time step of the simulation.

In Figure 6.25, the entries of the displacement vector $\boldsymbol{u} = (u_x, u_y, u_z)$ are presented. The resulting deformation is illustrated in Figure 6.26a.

**Creep during heating and austenitisation (S2)**  Considering higher temperatures, one has to take possible phase transitions into account, too. The implemented model for the arising phase transformations was presented in Section 2.3. Here, the given temperature increases to $\theta_{max} = 850°C$ and is kept constant afterwards (see Figure 6.23). Figure 6.22 presents the temperature in the inside of the beam during the simulation at $t = 170s$. The creep strain is calculated if $\theta \geq 300°C$ holds (cf. Remark 6.4.1). Figure 6.27 shows the evolution of the austenite phase fraction at different time steps in the interior of the simulated geometry.

More results of a simulation taking creep and phase transformations into account are presented in the following Section 6.4.2, see Figures 6.33–6.35.

Figure 6.22: Temperature during heating (S1), cut through beam.

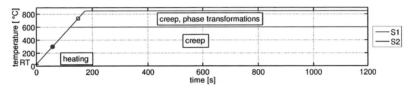

Figure 6.23: Given external temperature course in simulation (S1,S2), $\theta_{maxS1} = 600°C$, $\theta_{maxS2} = 850°C$. While $\theta \geq 300°C$, the creep strain is calculated (marked by blue circle). The beginning of the austenite formation is marked by the red square.

**Creep under a constant temperature (S3)**   Here, the temperature is kept constant at $\theta = 600°C$. Results of the simulation are presented in the following Section 6.4.2, see Figures 6.36–6.37.

**Creep during heating and slow cooling (S4)**   The external temperature given in the simulation increases linearly up to $700°C$, is kept constant and then decreases to room temperature, see Figure 6.28.

Figure 6.29 represents the resulting entries of the creep strain tensor $\varepsilon_c$ at the end of the simulation. In Figure 6.30, the corresponding entries of the displacement vector are shown. The resulting deformation of the beam is illustrated in Figure 6.26b.

**Creep during heating, austenitisation and slow cooling (S5)**   In this simulation, the temperature rises to $940°C$ (see Figure 6.31). Thus we have to take phase transformations into account.

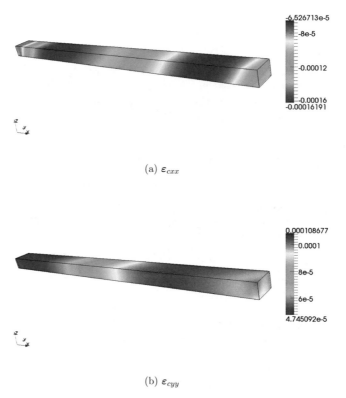

(a) $\varepsilon_{cxx}$

(b) $\varepsilon_{cyy}$

Figure 6.24: Creep during heating: Entries of creep strain tensor $\boldsymbol{\varepsilon}_c$ (as denoted in (6.4.1)) at $t = 1200s$, $\theta = 600°\text{C}$.

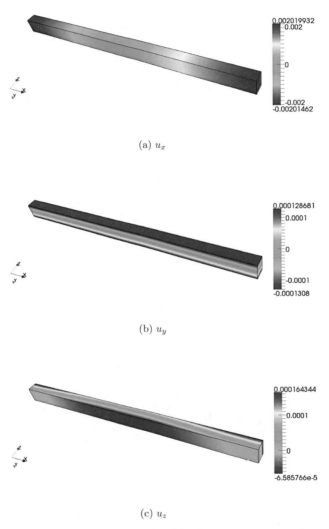

(a) $u_x$

(b) $u_y$

(c) $u_z$

Figure 6.25: Creep during heating: Entries of displacement vector $\boldsymbol{u} = (u_x, u_y, u_z)$ at $t = 1200s$, $\theta = 600°$C.

(a) Simulation S1: deformation at $t = 1200s$, $\theta = 600°$C

(b) Simulation S4: deformation at $t = 7950s$, $\theta = $ RT

Figure 6.26: Simulated deformation during different simulations with outline of undeformed geometry (deformation graphically increased by factor $2 \cdot 10^2$).

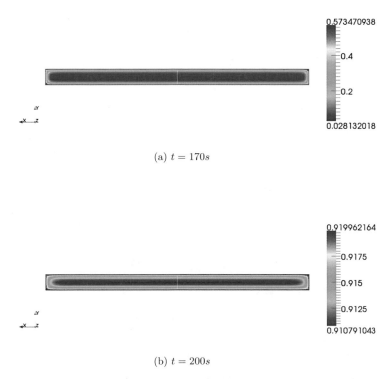

(a) $t = 170s$

(b) $t = 200s$

Figure 6.27: Simulation S2: Austenite phase fractions during heating at different time steps (cf. Figure 6.23), cut through beam.

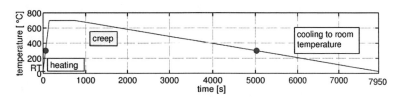

Figure 6.28: Simulation S4: Given external temperature in simulation, $\theta_{max} = 700°C$. While $\theta \geq 300°C$, the creep strain is calculated (marked by blue circles).

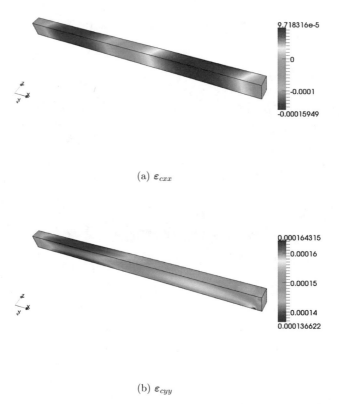

(a) $\varepsilon_{cxx}$

(b) $\varepsilon_{cyy}$

Figure 6.29: Creep during heating: Entries of creep strain tensor $\boldsymbol{\varepsilon}_c$ (as denoted in (6.4.1)) at $t = 7950s$, $\theta = $ RT (after heating to $\theta = 700°$C, see Figure 6.28).

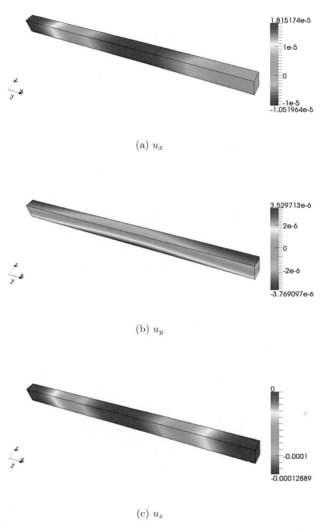

(a) $u_x$

(b) $u_y$

(c) $u_z$

Figure 6.30: Creep during heating: Entries of displacement vector $\boldsymbol{u} = (u_x, u_y, u_z)$ at $t = 7950s$, $\theta = \mathrm{RT}$ (after heating to $\theta = 700°C$, see Figure 6.28).

**Remark 6.4.1.** *In the presented simulations, the calculation of the creep strain starts when a specific temperature is reached (e.g., if $\theta \geq 300°C$ holds).*

*The reason for this is that the used material parameters were obtained within certain temperature regions (see Figure 6.2), cf. Section 6.1.*

## 6.4.2 Evolution in single nodes of geometry

Next, we investigate the material behaviour with respect to time in specific points of the geometry. Figure 6.32 shows the considered selection of twenty nodes in the mesh.

**Creep during heating and austenitisation (S2)** We give the external temperature in the simulation as plotted in Figure 6.23 where $\theta_{max} = 850°$C. The calculation of the creep strain starts if $\theta \geq 300°$C (cf. Remark 6.4.1).

The results for the selected nodes are presented in Figures 6.33–6.35: In Figure 6.33, the corresponding values of temperature as well as the phase fractions are plotted. In Figure 6.34, the resulting displacement $u_z$ together with the norm of the creep strain tensor is shown. Figure 6.35 summarises the entries fo the displacement vector $\boldsymbol{u} = (u_x, u_y, u_z)$.

**Creep under a constant temperature (S3)** We start with the heated workpiece taking the initial thermal expansion into account. The temperature is kept constant at $\theta = 600°$C. The resulting values of the norm of the creep strain tensor as well as the displacement $u_z$ are presented in Figure 6.36. The results for the single entries of $\boldsymbol{u}$ are shown in Figure 6.37.

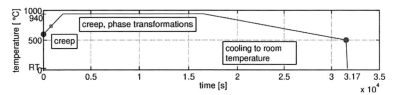

Figure 6.31: Simulation S5: Evolution of external temperature in simulation, $\theta_{max} = 940°C$, $\theta_{end} = RT$. The creep strain is calculated while $\theta \geq 500°C$ (marked by blue circles). The square marks the start temperature for the formation of austenite.

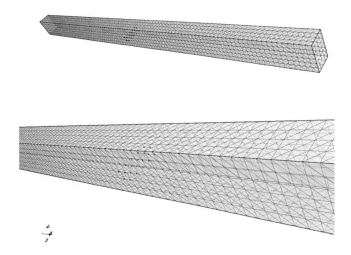

Figure 6.32: Selection of nodes.

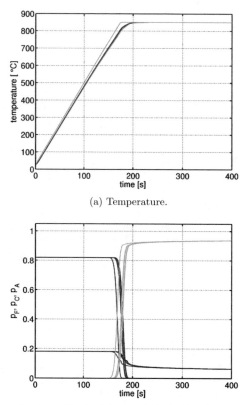

(a) Temperature.

(b) Phase fractions of ferrite (blue), carbide (red) and austenite (cyan).

Figure 6.33: Phase transformations: Temperature and phase fractions during heating and austenitisation.

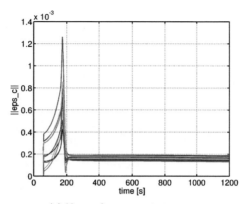

(a) Norm of creep strain tensor $\varepsilon_c$.

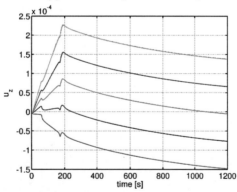

(b) Displacement $\boldsymbol{u}$ in $z$-direction.

Figure 6.34: Values in single nodes: Creep strain and displacement under varying temperature, $\theta_0 = RT$, $\theta_{max} = 850°C$. The value of the creep strain is calculated when $\theta \geq 300°C$.

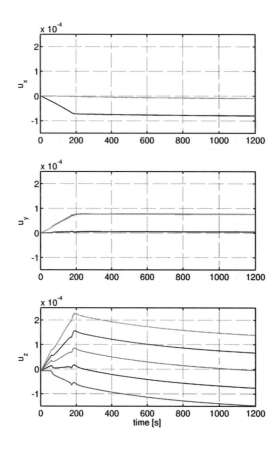

Figure 6.35: Values in single nodes ($\theta_0 = RT$, $\theta_{max} = 850°C$): Entries of displacement vector $\boldsymbol{u} = (u_x, u_y, u_z)$.

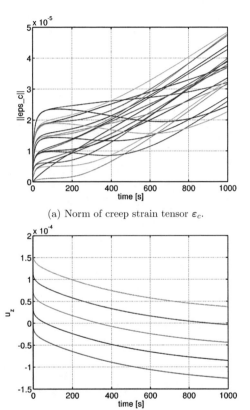

(a) Norm of creep strain tensor $\varepsilon_c$.

(b) Displacement $\boldsymbol{u}$ in $z$-direction.

Figure 6.36: Values in single nodes (simulation under constant temperature $\theta = 600°C$).

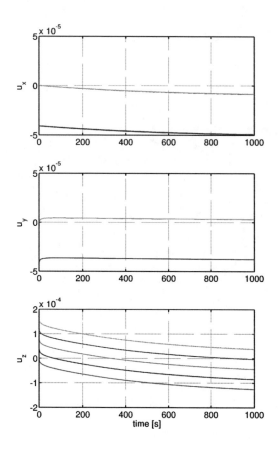

Figure 6.37: Values in single nodes (simulation under constant temperature $\theta = 600°C$): Entries of displacement vector $\boldsymbol{u} = (u_x, u_y, u_z)$.

### 6.4.3 Validation by experimental data

In this section, we compare the calculated deformation to experimental data. The experimental setup is as described above, see Figure 6.18. The temperatures during the experiments are summarised in Table 6.7[1]. In the second experimental row, the steel beam is heated up to 940°C such that phase transformations occur during heating and cooling down.

After the performed experiment, the displacement of the steel beam is measured along three horizontal lines. This is done at three different faces of the beam. The measured data consists of the displacement in the orthogonal direction of the considered beam's face. The resulting data for an experiment with $\theta_{max} = 700°C$ is presented in Figure 6.38. In Figure 6.39 the three measured displacements are plotted each over the beam's length.

In order to compare the simulation results with experimental data, we performed a simulation under a similar external temperature course as during the experiment. The given temperature in the simulation is plotted in Figure 6.28.

We consider the simulation results in the last time step in specific nodes of the geometry. The selection of nodes is shown in Figure 6.40. Thus, we are able to consider the displacement with respect to the beam's length.

Figure 6.41 shows the simulated and measured displacement of the lower face of the beam in $z$-direction. Here, '$z$-direction' refers to the direction of the corresponding axis as illustrated in Figure 6.20.

**Remark 6.4.2** (Experiment and simulation). *Comparing the resulting deformation of the experiment and of the simulation there are some deviations. The reason for this are possibly due to the implemented support (elements of the geometry are fixed in z-direction at the support).*

*Furthermore, the material data used in the simulation, do not correspond to the material of the steel beam in the experiment as we did not have information about that specific material. The material used in the experiments is a C45 (SAE 1045) steel.*

---

[1]The experimental data was provided by the Institute of Materials Science Bremen (IWT).

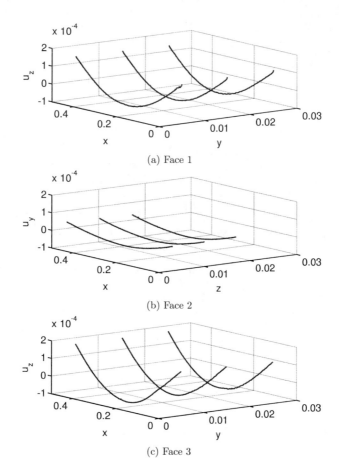

(a) Face 1

(b) Face 2

(c) Face 3

Figure 6.38: Creep experiments using steel beams: measured displacement over beam length at different faces of the beam after the performed experiment ($\theta_{max} = 700°C$).

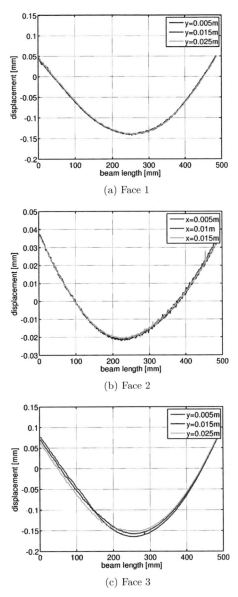

(a) Face 1

(b) Face 2

(c) Face 3

Figure 6.39: Creep experiments using steel beams: Measured displacements over beam length.

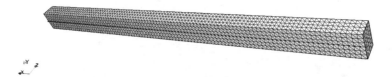

Figure 6.40: Selection of nodes.

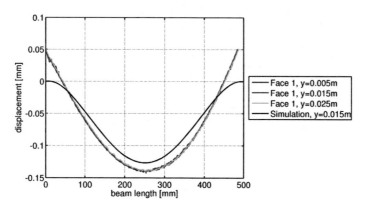

Figure 6.41: Comparison between experiment and simulation: displacement over beam length.

### 6.4.4 3D Simulation of a steel ring

The model presented in Chapter 5 was used to simulate the heating of an SAE 52100 steel ring. The dimensions of the implemented geometry are as follows: The simulated ring has the outer radius $r_{out} = 72.5$ mm, the inner radius $r_{in} = 66.5$ mm and a height of $h = 26$ mm.

The dimensions correspond to the steel ring shown in Figure 6.19 located on a support in the gas nozzle field. In the simulations, we assume a support at four positions of the ring. Thereby, we are able to restrict the simulated geometry to a quarter in circumference and make use of symmetric boundary conditions. Figure 6.42 shows the implemented geometry with the mesh.

We consider a simulation where the workpiece is heated up to 700°C. The resulting distribution of the temperature in the ring is shown in Figure 6.43. Figure 6.44 and 6.45 present the calculated entries of the creep strain tensor $\varepsilon_c$ and of the displacement $\boldsymbol{u}$, respectively. Figure 6.46 demonstrates the expansion and the deformation during heating.

Figure 6.42: Simulated ring with mesh.

Figure 6.43: Temperature distribution during simulation.

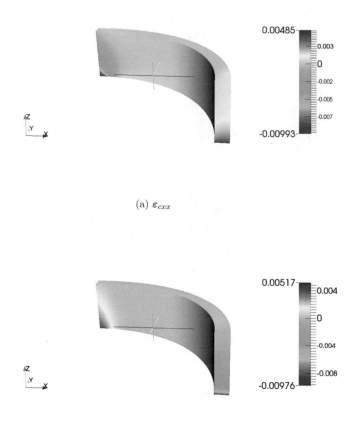

(a) $\varepsilon_{cxx}$

(b) $\varepsilon_{cyy}$

Figure 6.44: Creep during heating: Entries of creep strain tensor $\varepsilon_c$ (as denoted in (6.4.1)) at $t = 4.2s$.

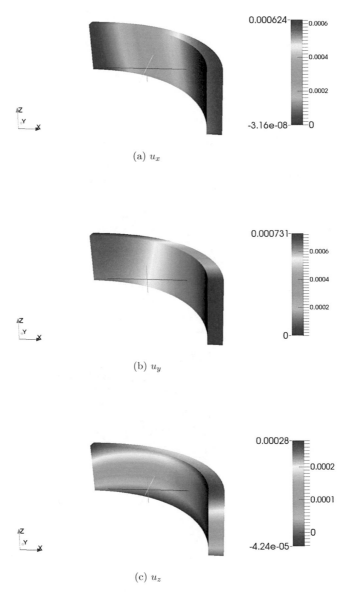

(a) $u_x$

(b) $u_y$

(c) $u_z$

Figure 6.45: Creep during heating: Entries of displacement vector $\boldsymbol{u} = (u_x, u_y, u_z)$ at $t = 4.2s$.

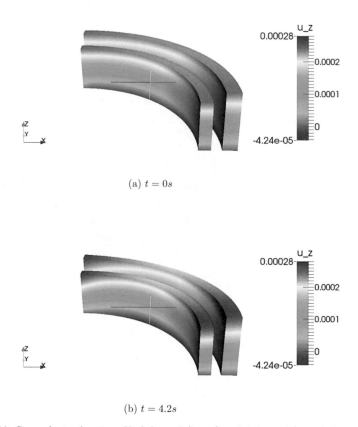

(a) $t = 0s$

(b) $t = 4.2s$

Figure 6.46: Creep during heating: Undeformed (inner) and deformed (outer) ring
at different time steps (the deformation is graphically increased by a
factor of $2 \cdot 10^2$).

# 7 Outlook

In this thesis we provided the complete mathematical model for the material behaviour of steel during heating and austenitisation, developed a numerical solution scheme and presented results from the implemented model equations in a one-dimensional as well as in a three-dimensional setting. The obtained simulation results showed that the models are in good agreement with reality.

Considering inelastic material behaviour, we presented an alternative modelling approach by multi-mechanism models. In particular, we focused on two-mechanism models and provided applications for creep as well as for creep and TRIP.

By the developed numerical solution scheme we were able to implement and to simulate the material behaviour of steel in three dimensions. This enabled us to conduct 3D simulations of the heating process of a workpiece under realistic conditions. We validated the model approach by means of data from workpiece experiments. Altogether, the 3D simulations showed that the model is in good agreement with reality.

The outcome of this thesis forms a basis for future investigations. The next step would be the simulation of the entire heat treatment process including heating, austenitisation as well as quenching with arising phase transformation and (coupled) inelastic phenomena. Hereby, one is able to observe the influence of the phenomena during heating and austenitisation on the material behaviour afterwards. This results finally in an improved prediction of the distortion of a steel workpiece during heat treatment.

Additional tasks for future work are:

- further validations of the implemented 3D model by means of experimental data

- different meshing strategies and possibly adaptive methods

- involving the simultaneous occurrence of creep and TRIP in the 3D simulation of heating and austenitisation

- alternative model approaches for the inelastic strains, e.g. via multi-mechanism models

The results of the parameter identification as well as of the model verification for the one-dimensional case showed that the creep behaviour of both states of the material are in good accordance with experimental data. However, the creep curves calculated by the stress-driven mode differ from experiments. A cause for this may be the composite structure of AGC requiring a more complex creep model.

Special difficulties arise when investigating TRIP during austenitisation which occurs together with creep. The inelastic strain of the phase mixture is assumed as the sum of the TRIP strain and the creep strain. The latter is calculated using a mixture rule and extrapolation which might be a source of errors.

In future, additional austenitisation experiments investigating creep and TRIP will be necessary in order to confirm and improve parameter values.

One of the novelties in this thesis is the investigation of the coupling of several inelastic material phenomena as creep and TRIP during heating. The approach by multi-mechanism models is relatively new and there are not yet many results in this field regarding the application and the simulation of specific models. Here, we presented different two-mechanism model for creep as well as for creep and TRIP.

A next step would be the implementation of the model equations, together with a parameter identification and the validation of material laws by experimental data. Furthermore, future prospects could consist in different model approaches such as multi mechanism models with more than two mechanisms. Besides uniaxial experiments, also the modelling of biaxial experiments with a torsion of the steel specimen can be considered.

# A Semi-implicit version of numerical algorithm in Section 5.4

Here, we present an alternative approach to the one presented in Chapter 5 in Box 5.4.1. Instead of solving the problem implicitly, we present an alternative approach taking some values from the former time step $t_{n-1}$.

---

### A.0.1. Numerical algorithm for the calculation of inelastic quantities

Start with initial values $\theta^0, p^0, \boldsymbol{u}^0, \boldsymbol{\sigma}^0, \varepsilon^0, \varepsilon_c^0, \boldsymbol{X}^0$. Compute the current quantities at time $t_n$. All quantities for $t_{n-1}$ are known.

The current temperature $\theta^n$, the phase fractions $p^n$ as well as the deformation $\boldsymbol{u}^n$ are given by step I) of the algorithm in Box 5.2.1.

### Step II): Calculate inelastic quantities $\varepsilon_c^n, \boldsymbol{X}^n, \boldsymbol{\sigma}^n$

1. Calculate (the deviator of) the trial stress $\boldsymbol{\sigma}_t^{*n}$

2. Solve for $\varepsilon_c^n$ via semi-implicit approach

3. Update inelastic quantities, i.e. $\varepsilon_c^n$, $s_c^n$, $\boldsymbol{X}_c^n$

4. Correct stress tensor $\boldsymbol{\sigma}^n$ using the updated value of $\varepsilon_c^n$

---

### Semi-implicit approach

We define the effective stress as

$$\xi^n := \boldsymbol{\sigma}^{*n} - \boldsymbol{X}_c^{*n} \ . \tag{A.0.1}$$

Together with (5.4.11) we obtain

$$\xi^n = \boldsymbol{\sigma}^{*n} - \boldsymbol{X}_c^{*n} = \boldsymbol{\sigma}_t^{*n} - \boldsymbol{X}_c^{*n} - 2\mu\tau_n(\dot{\varepsilon}_c)^n \tag{A.0.2}$$

We use the following *semi-implicit approach* for the discretisation of the back stress taking the value of $\dot{\boldsymbol{\varepsilon}}_c$ from the former time step:

$$\boldsymbol{X}_c^n = \boldsymbol{X}_c^{n-1} + \tau_n(\dot{\boldsymbol{X}}_c)^n$$

$$= \boldsymbol{X}_c^{n-1} + \tau_n \frac{2}{3} c_c(\theta^n) \, (\dot{\boldsymbol{\varepsilon}}_c)^{n-1} - \tau_n b_c(\theta^n) \, \boldsymbol{X}_c^n \, (\dot{s}_c)^n + \Delta c_c(\theta^n) \boldsymbol{X}_c^n \qquad \text{(A.0.3)}$$

where $\Delta c_c(\theta^n) = \frac{c_c(\theta^n) - c_c(\theta^{n-1})}{c_c(\theta^n)}$. For the discretisation of $\dot{s}_c$ (cf. (2.2.31)) we take the following approach

$$(\dot{s}_c)^n = \sqrt{\frac{2}{3}} \|(\dot{\boldsymbol{\varepsilon}}_c)^{n-1}\| \, . \qquad \text{(A.0.4)}$$

Thus, equation (A.0.3) can be transformed into

$$\boldsymbol{X}_c^n = \boldsymbol{X}_c^{n-1} + \tau_n \frac{2}{3} c_c(\theta^n) \, (\dot{\boldsymbol{\varepsilon}}_c)^{n-1} - \tau_n b_c(\theta^n) \, \boldsymbol{X}_c^n \sqrt{\frac{2}{3}} \|(\dot{\boldsymbol{\varepsilon}}_c)^{n-1}\| + \Delta c_c(\theta^n) \boldsymbol{X}_c^n \, .$$
$$\text{(A.0.5)}$$

Analogously to the discretisation of the back stress, we also use a semi-implicit approach for the effective stress $\xi^n$ by taking the value of $(\dot{\boldsymbol{\varepsilon}}_c)$ from the former time step. Now, after having determined the current value of the back stress $\boldsymbol{X}^n$ we are able to obtain:

$$\xi^n = \boldsymbol{\sigma}_t^{*n} - \boldsymbol{X}_c^{*n} - 2\mu\tau_n(\dot{\boldsymbol{\varepsilon}}_c)^{n-1} \, . \qquad \text{(A.0.6)}$$

After that, we can calculate the current value of the time derivative of the creep strain $(\dot{\boldsymbol{\varepsilon}}_c)^n$ by the equation

$$(\dot{\boldsymbol{\varepsilon}}_c)^n = \frac{3}{2} A_n \left( \sqrt{\frac{3}{2}} \|\xi^n\| \right)^{m_n - 1} \xi^n (s_c^{n-1} + \tau_n \sqrt{\frac{2}{3}} \|(\dot{\boldsymbol{\varepsilon}}_c)^{n-1}\|)^k \qquad \text{(A.0.7)}$$

Finally, we can calculate the corrected stress tensor by

$$\boldsymbol{\sigma}^n = \boldsymbol{\sigma}_t^{*n} + (\lambda + \frac{2}{3}\mu)\text{tr}(\boldsymbol{\varepsilon}^n)\boldsymbol{I} - 2\mu\tau_n(\dot{\boldsymbol{\varepsilon}}_c)^n \, , \qquad \text{(A.0.8)}$$

and update the values

$$\boldsymbol{\varepsilon}_c^n = \boldsymbol{\varepsilon}_c^{n-1} + \tau_n(\dot{\boldsymbol{\varepsilon}}_c)^n \qquad \text{(A.0.9)}$$

and

$$s_c^n = s_c^{n-1} + \tau_n(\dot{s}_c)^n = s_c^{n-1} + \tau_n \sqrt{\frac{2}{3}} \|(\dot{\boldsymbol{\varepsilon}}_c)^n\| \, . \qquad \text{(A.0.10)}$$

**Step 2b: Inelastic strain: Creep strain, back stress**

In the next step, we have a look at the discretisation of (2.2.30) and (2.2.34). In the following, we set $l = 1$ and $D_c = 1$.

We define the effective stress as

$$\xi^n := \boldsymbol{\sigma}^{*n} - \boldsymbol{X}_c^{*n} .\tag{A.0.11}$$

Together with (5.4.11) we obtain

$$\xi^n = \boldsymbol{\sigma}^{*n} - \boldsymbol{X}_c^{*n} = \boldsymbol{\sigma}_t^{*n} - \boldsymbol{X}_c^{*n} - 2\mu\tau_n(\dot{\boldsymbol{\varepsilon}}_c)^n \tag{A.0.12}$$

We use the following *semi-implicit approach* for the discretization of the back stress taking the value of $\dot{\boldsymbol{\varepsilon}}_c$ from the former time step:

$$\boldsymbol{X}_c^n = \boldsymbol{X}_c^{n-1} + \tau_n(\dot{\boldsymbol{X}}_c)^n$$
$$= \boldsymbol{X}_c^{n-1} + \tau_n\frac{2}{3}c_c(\theta^n)\,(\dot{\boldsymbol{\varepsilon}}_c)^{n-1} - \tau_n b_c(\theta^n)\,\boldsymbol{X}_c^n\,(\dot{s}_c)^n + \Delta c_c(\theta^n)\boldsymbol{X}_c^n \tag{A.0.13}$$

where $\Delta c_c(\theta^n) = \frac{c_c(\theta^n) - c_c(\theta^{n-1})}{c_c(\theta^n)}$. For the discretisation of $\dot{s}_c$ (cf. (2.2.31)) we take the following approach

$$(\dot{s}_c)^n = \sqrt{\frac{2}{3}}\|(\dot{\boldsymbol{\varepsilon}}_c)^{n-1}\| .\tag{A.0.14}$$

Thus, equation (A.0.3) can be transformed into

$$\boldsymbol{X}_c^n = \boldsymbol{X}_c^{n-1} + \tau_n\frac{2}{3}c_c(\theta^n)\,(\dot{\boldsymbol{\varepsilon}}_c)^{n-1} - \tau_n b_c(\theta^n)\,\boldsymbol{X}_c^n\,\sqrt{\frac{2}{3}}\|(\dot{\boldsymbol{\varepsilon}}_c)^{n-1}\| + \Delta c_c(\theta^n)\boldsymbol{X}_c^n .$$
$$\tag{A.0.15}$$

Analogously to the discretization of the back stress, we also use a semi-implicit approach for the effective stress $\xi^n$ by taking the value of $(\dot{\boldsymbol{\varepsilon}}_c)$ from the former time step. Now, after having determined the current value of the back stress $\boldsymbol{X}^n$ we are able to obtain:

$$\xi^n = \boldsymbol{\sigma}_t^{*n} - \boldsymbol{X}_c^{*n} - 2\mu\tau_n(\dot{\boldsymbol{\varepsilon}}_c)^{n-1} .\tag{A.0.16}$$

After that, we can calculate the current value of the time derivative of the creep strain $(\dot{\boldsymbol{\varepsilon}}_c)^n$ by the equation

$$(\dot{\boldsymbol{\varepsilon}}_c)^n = \frac{3}{2}A_n\left(\sqrt{\frac{3}{2}}\|\xi^n\|\right)^{m_n-1}\xi^n(s_c^{n-1} + \tau_n\sqrt{\frac{2}{3}}\|(\dot{\boldsymbol{\varepsilon}}_c)^{n-1}\|)^k \tag{A.0.17}$$

Finally, we can calculate the corrected stress tensor by

$$\boldsymbol{\sigma}^n = \boldsymbol{\sigma}_t^{*n} + (\lambda + \frac{2}{3}\mu)\mathrm{tr}(\boldsymbol{\varepsilon}^n)\boldsymbol{I} - 2\mu\tau_n(\dot{\boldsymbol{\varepsilon}}_c)^n , \tag{A.0.18}$$

and update the values

$$\boldsymbol{\varepsilon}_c^n = \boldsymbol{\varepsilon}_c^{n-1} + \tau_n(\dot{\boldsymbol{\varepsilon}}_c)^n \tag{A.0.19}$$

and

$$s_c^n = s_c^{n-1} + \tau_n(\dot{s}_c)^n = s_c^{n-1} + \tau_n\sqrt{\frac{2}{3}}\|(\dot{\boldsymbol{\varepsilon}}_c)^n\| .\tag{A.0.20}$$

# Bibliography

Acht, C., Dalgic, M., Frerichs, F., Hunkel, M., Irretier, A., Lübben, T., and Surm, H. (2008a). Ermittlung der Materialdaten zur Simulation des Durchhärtens von Komponenten aus 100Cr6 - Teil 1. *HTM Journal of Heat Treatment and Materials*, 63:234 – 244.

Acht, C., Dalgic, M., Frerichs, F., Hunkel, M., Irretier, A., Lübben, T., and Surm, H. (2008b). Ermittlung der Materialdaten zur Simulation des Durchhärtens von Komponenten aus 100Cr6 - Teil 2. *HTM Journal of Heat Treatment and Materials*, 63:362 – 371.

Ahrens, U., Besserdich, G., and Maier, H. J. (2000). Spannungsabhängiges bainitisches und martensitisches Umwandlungsverhalten eines niedrig legierten Stahl. *HTM Haerterei-Techn. Mitt.*, 55:329 – 338.

Altenbach, H. (2012). *Kontinuumsmechanik*. Springer-Verlag.

Altenbach, H. and Altenbach, J. (1994). *Einführung in die Kontinuumsmechanik*. Teubner-Verlag.

Arya, V. K. and Kaufman, A. (1989). Finite element implementation of Robinson's unified viscoplastic model and its application to some uniaxial and multiaxial problems. *Eng. Comput.*, 6:237 – 247.

Berns, H. and Theisen, W. (2008). *Eisenwerkstoffe - Stahl und Gusseisen*. Springer-Verlag, Berlin, Heidelberg.

Bertram, A. and Glüge, R. (2013). *Festkörpermechanik*. Otto-von-Guericke-Universität Magdeburg.

Besson, J., Cailletaud, G., Chaboche, J.-L., and Forest, S. (2001). *Mécanique non linéaire des matériaux*. Hermes Science Europe Ltd.

Betten, J. (2001). *Kontinuumsmechanik*. Springer-Verlag, Berlin.

Betten, J. (2002). *Creep Mechanics*. Engineering online library. Springer-Verlag, Berlin.

Boettcher, S. (2012). *Modelling, analysis and simulation of thermo-elasto-plasticity with phase transitions in steel*. PhD thesis, Universität Bremen, Berlin.

Boettcher, S., Böhm, M., and Wolff, M. (2015). Well-posedness of a thermo-elasto-plastic problem with phase transitions in TRIP steels under mixed boundary conditions. *ZAMM - Journal of Applied Mathematics and Mechanics / Zeitschrift für Angewandte Mathematik und Mechanik.*

Bökenheide, S., Montalvo Urquizo, J., and Wolff, M. (2012a). Modelling of creep and TRIP during heating and austenitisation. European Congress on Computational Methods in Applied Sciences and Engineering (ECCOMAS) 2012, Wien, Austria.

Bökenheide, S. and Wolff, M. (2012). Comparison of different approaches to verify creep behaviour of 100Cr6 steel. *Computational Materials Science*, 64:34–37.

Bökenheide, S., Wolff, M., Dalgic, M., Lammers, D., and Linke, T. (2011). Creep, phase transformations and transformation-induced plasticity of 100cr6 steel during heating. IDE 2011, Bremen, Germany.

Bökenheide, S., Wolff, M., Dalgic, M., Lammers, D., and Linke, T. (2012b). Creep, phase transformations and transformation-induced plasticity of 100Cr6 steel during heating. *Mat.-wiss. u. Werkstofftech.*, 43(1-2):143–149.

Braess, D. (1992). *Finite Elemente: Theorie, schnelle Löser und Anwendungen in der Elastizitätstheorie.* Springer Lehrbuch. Springer, Berlin u.a.

Cailletaud, G. and Saï, K. (1995). Study of plastic/viscoplastic models with various inelastic mechanisms. *Int. J. of Plast.*, 11:991 – 1005.

Chaboche, J. (2008). A review of some plasticity and viscoplasticity constitutive theories. *International Journal of Plasticity*, 24:1642 – 1693.

Ciarlet, P. (2002). *The Finite Element Method for Elliptic Problems.* Classics in Applied Mathematics. Society for Industrial and Applied Mathematics.

Dalgic, M., Irretier, A., and Zoch, H.-W. (2009). Stress-strain behaviour of the case hardening steel 20MnCr5 regarding different microstructures. *Mat.-wiss. u. Werkstofftech.*, 40:448 – 453.

Dalgic, M. and Löwisch, G. (2006). Transformation plasticity at different phase transformation of a bearing steel. *Materialwissenschaften und Werkstofftechnik*, 37:122 – 127.

Dautray, R. and Lions, J. L. (1992). *Evolution Problems I*, volume 5 of *Mathematical Analysis and Numerical Methods for Science and Technology.* Springer.

de Souza Neto, E., Perić, D., and Owen, D. (2008). *Computational Methods for plasticity.* Wiley.

Feynman, R. P., Leighton, R. B., and Sands, M. (1991). *Vorlesungen über Physik, Bd II.* Oldenbourg Verlag, München Wien.

Fischer, F. D., Sun, Q. P., and Tanaka, K. (1996). Transformation-induced plasticity (TRIP). *Appl. Mech. Rev.*, 49:317 – 364.

Han, W. and Reddy, D. B. (1999). *Plasticity - mathematical theory and numerical analysis*. Springer.

Haupt, P. (2002). *Continuum Mechanics and Theory of Materials*. Springer-Verlag.

Helm, D. and Haupt, P. (2003). Shape memory behavior: modeling within continuum thermomechanics. *Int. J. of Solids and Structures*, 40:827.

Horstmann, D. (1992). *Das Zustandsschaubild Eisen-Kohlenstoff und die Grundlagen der Wärmebehandlung der Eisen-Kohlenstoff-Legierungen*. Verlag Stahleisen.

Ilschner, B. and Singer, R. F. (2005). *Werkstoffwissenschaften und Fertigungstechnik*. Springer.

Knabner, P. and Angermann, L. (2000). *Numerik partieller Differentialgleichungen*. Springer.

Kröger, N. H. (2013). *Multi-Mechanism Models - Theory and Applications*. PhD thesis, Universität Bremen.

Leblond, J. B. (1989). Mathematical modelling of transformation plasticity in steels. II: Coupling with strain hardening phenomena. *Int. J. of Plast.*, 5:573 – 591.

Lemaitre, J. and Chaboche, J.-L. (1990). *Mechanics of solid materials*. Cambridge University Press, Cambridge [u.a.].

Mahnken, R. (2004). *Identification of Material Parameters for Constitutive Equations*, volume 2 of *Encyclopedia of Computational Mechanics*, chapter 19, pages 637 – 655. John Wiley & Sons, Ltd.

Mahnken, R., Schneidt, A., Antretter, T., Ehlenbröker, U., and Wolff, M. (2015). Multi-scale modeling of bainitic phase transformation in multi-variant polycrystalline low alloy steels. *International Journal of Solids and Structures*, 54:156 – 171.

Mahnken, R. and Stein, E. (1996). Parameter identification for viscoplastic models based on analytical derivatives of a least-squares functional and stability investigations. *Int. J. of Plast.*, 12:451 – 479.

Mahnken, R., Wolff, M., Schneidt, A., and Böhm, M. (2012). Multi-phase transformations at large strains – thermodynamic framework and simulation. *International Journal of Plasticity*, 39:1 – 26.

Maugin, G. (1992). *The Thermomechanics of Plasticity and Fracture*. Cambridge University Press.

Naumenko, K. and Altenbach, H. (2007). *Modeling of Creep for Structural Analysis*. Springer-Verlag.

Palmov, V. A. (1998). *Vibrations of elasto-plastic bodies*. Springer-Verlag.

Porter, D. A. and Easterling, K. E. (1992). *Phase Transformations in Metals and Alloys*. CRC Press.

Regrain, C., Laiarinandrasana, L., Toillon, S., and Saï, K. (2009). Multi-mechanism models for semi-crystalline polymer: Constitutive relations and finite element implementation. *Int. J. Plasticity*, 25:1253–1279.

Reichelt, G. (1982). *Beitrag zum Austenitisierungsprozeß der Stähle*. PhD thesis, TU Berlin.

Sai, K. (1993). *Modèles à grand nombre de variables internes et méthodes numériques associées*. PhD thesis, National Ecole des Mines de Paris.

Saï, K. (2011). Multi-mechanism models: Present state and future trends. *International Journal of Plasticity*, 27:250 – 281.

Saï, K. and Cailletaud, G. (2007). Multi-mechanism models for the description of ratcheting: Effect of the scale transition rule and of the coupling between hardening variables. *Int. J. of Plast.*, 23:1589 – 1617.

Saï, K., Laiarinandrasana, L., Naceur, I. B., Besson, J., Jeridi, M., and Cailletaud, G. (2011). Multi-mechanism damage-plasticity model for semi-crystalline polymer: Creep damage of notched specimen of pa6. *Materials Science and Engineering A*, 528:1087–1093.

Schmidt, A. and Siebert, K. (2005). *Design of Adaptive Finite Elemente Software - The Finite Element Toolbox ALBERTA, LNCSE Series 42*. Springer Verlag, Berlin, Heidelberg.

Schmidt, A., Wolff, M., and Böhm, M. (2003). Numerische Untersuchungen für ein Modell des Materialverhaltens mit Umwandlungsplastizität und Phasenumwandlungen beim Stahl 100Cr6. Technical Report 03-13, Berichte aus der Technomathematik, FB 3, Universität Bremen.

Seidel, W. (1999). *Werkstofftechnik*. Carl Hanser Verlag.

Simo, J. C. and Hughes, T. J. R. (1998). *Computational inelasticity*. Springer-Verlag.

Suhr, B. (2010). *Simulation of steel quenching with interaction of classical plasticity and TRIP - numerical methods and model comparison*. PhD thesis, Universität Bremen.

Surm, H., Kessler, O., Hoffmann, F., and Zoch, H.-W. (2008). Modelling of austenitising with non-constant heating rate in hypereutectoid steels. *International Journal of Microstructure and Materials Properties*, 3(1):35 – 48.

Taleb, L. and Cailletaud, G. (2010). An updated version of the multimechanism model for cyclic plasticity. *Int. Journal of Plast.*, 26(6):859 – 874.

Wolff, M., Boettcher, S., and Böhm, M. (2007). Phase transformations in steel in the multi-phase case - general modelling and parameter identification. Technical Report 07-02, Berichte aus der Technomathematik, FB 3, Universität Bremen.

Wolff, M. and Böhm, M. (2002a). Phasenumwandlungen und Umwandlungsplastizität bei Stählen im Konzept der Thermoelasto-Plastizität. Technical Report 02-05, Berichte aus der Technomathematik, FB 3, Universität Bremen.

Wolff, M. and Böhm, M. (2002b). Zur Modellierung der Thermoelasto-Plastizität mit Phasenumwandlungen bei Stählen sowie der Umwandlungsplastizität. Technical Report 02-01, Berichte aus der Technomathematik, FB 3, Universität Bremen.

Wolff, M. and Böhm, M. (2010). Two-mechanism models and modelling of creep. In Mikhlin, J. and Perepelkin, M., editors, *Proc. of the 3rd International Conference on Nonlinear Dynamics*, Kharkov, Ukraine. National Technical University "Kharkov Polytechnic Institute". ISBN 978-966-8230-42-4.

Wolff, M., Böhm, M., Bänsch, E., and Davis, D. (2000). Modellierung der Abkühlung von Stahlbrammen. Technical Report 00-07, Berichte aus der Technomathematik, FB 3, Universität Bremen.

Wolff, M., Böhm, M., Bökenheide, S., and Dalgic, M. (2011a). Some recent developments in modelling of heat-treatment phenomena in steel within the collaborative research centre SFB 570 'Distortion Engineering'. IDE 2011, Bremen, Germany.

Wolff, M., Böhm, M., Bökenheide, S., and Dalgic, M. (2012a). Some recent developments in modelling of heat-treatment phenomena in steel within the collaborative research centre SFB 570 'Distortion Engineering'. *Mat.-wiss. u. Werkstofftech.*, 43(1-2):136–142.

Wolff, M., Böhm, M., Bökenheide, S., and Kröger, N. H. (2012b). Two-mechanism approach in thermo-viscoelasticity with internal variables. *Technische Mechanik*, 32(2):608–621.

Wolff, M., Böhm, M., Bökenheide, S., Lammers, D., and Linke, T. (2012c). An Implicit Algorithm to Verify Creep and TRIP Behavior of Steel Using Uniaxial Experiments. *ZAMM - Z. Angew. Math. Mech.*, 92(5):355–379.

Wolff, M., Böhm, M., and Dachkovski, S. (2003). Volumenanteile versus Massenanteile - der Dilatometerversuch aus der Sicht der Kontinuumsmechanik. Technical Report 03-01, Berichte aus der Technomathematik, FB 3, Universität Bremen.

Wolff, M., Böhm, M., and Helm, D. (2008). Material behavior of steel – modeling of complex phenomena and thermodynamic consistency. *Int. J. of Plast.*, 24:746–774.

Wolff, M., Böhm, M., Mahnken, R., and Suhr, B. (2011b). Implementation of an algorithm for general material behaviour of steel taking interaction between plasticity and transformation-induced plasticity into account. *Int. J. Numer. Meth. Engng*, 87:1183–1206.

Wolff, M., Böhm, M., and Suhr, B. (2009). Comparison of different approaches to transformation-induced plasticity in steel. *Materialwissenschaften und Werkstofftechnik*, 40(5-6):454–459.

Wolff, M., Böhm, M., and Taleb, L. (2010). Two-mechanism models with plastic mechanisms - modelling in continuum-mechanical framework. Technical Report 10-05, Berichte aus der Technomathematik, FB 3, Universität Bremen.

Wolff, M., Böhm, M., and Taleb, L. (2011c). Thermodynamic consistency of two-mechanism models in the non-isothermal case. *Technische Mechanik*, 31:58–80.

Wolff, M., Bökenheide, S., and Böhm, M. (2015). Some new extensions to multi-mechanism models for plastic and viscoplastic material behavior under small strains. *Continuum Mechanics and Thermodynamics*, doi: 10.1007/s00161-015-0418-5, to appear.

Wolff, M., Bökenheide, S., Schlasche, J., Büsing, D., Böhm, M., and Zoch, H.-W. (2013). An extended approach to multi-mechanism models in plasticity - theory and parameter identification. Technical Report 13-06, Berichte aus der Technomathematik, FB 3, Universität Bremen.

Wolff, M. and Taleb, L. (2008). Consistency for two multi-mechanism models in isothermal plasticity. *Int. J. of Plast.*, 24:2059 – 2083.

Yun, G. and Shang, S. (2011). A self-optimizing inverse analysis method for estimation of cyclic elasto-plasticity model parameters. *International Journal of Plasticity*, 27:576 – 595.

Zeidler, E. (1990). *Linear Monotone Operators*, volume II/A of *Nonlinear Functional Analysis and Its Applications*. Springer.